The Genesis of Living Forms

GROUNDWORKS

Series Editors:

Arne De Boever, California Institute of the Arts
Bill Ross, Staffordshire University
Jon Roffe, University of New South Wales
Ashley Woodward, University of Dundee

What are the hidden sources that determine the contemporary moment in continental thought? This series goes 'back to the source', publishing English translations of the hidden origins of our contemporary thought in order to better understand not only that thought, but also the world it seeks to understand. The series includes important French, German and Italian texts that form the lesser-known background to prominent work in contemporary continental philosophy. With an eye on the contemporary moment – on both world-historical events and critical trends – Groundworks seeks to recover foundational but forgotten texts and to produce a more profound engagement not only with the contemporary but also with the sources that have shaped it.

The Dialectic of Duration, by Gaston Bachelard
Translated and annotated by Mary McAllester Jones
Introduction by Cristina Chimisso

The Birth of Physics, by Michel Serres
Translated by David Webb and Bill Ross

The Genesis of Living Forms, by Raymond Ruyer
Translated by Jon Roffe and Nicholas B. de Weydenthal

The Genesis of Living Forms

Raymond Ruyer

Translated by Jon Roffe and
Nicholas B. de Weydenthal

ROWMAN &
LITTLEFIELD
———INTERNATIONAL
London • New York

Published by Rowman & Littlefield International Ltd
6 Tinworth Street, London, SE11 5AL, UK
www.rowmaninternational.com

Rowman & Littlefield International Ltd is an affiliate of Rowman & Littlefield
4501 Forbes Boulevard, Suite 200, Lanham, Maryland 20706, USA
With additional offices in Boulder, New York, Toronto (Canada), and Plymouth (UK)
www.rowman.com

Translation © 2020 Jon Roffe and Nicholas B. de Weydenthal

Originally published as *La Genèse des formes vivantes* by Editions Flammarion, 1958

All rights reserved. No part of this book may be reproduced in any form or by any electronic or mechanical means, including information storage and retrieval systems, without written permission from the publisher, except by a reviewer who may quote passages in a review.

British Library Cataloguing in Publication Data

A catalogue record for this book is available from the British Library

ISBN: HB 978-1-78660-087-5
ISBN: PB 978-1-78660-088-2

Library of Congress Cataloging-in-Publication Data Available

ISBN: 978-1-78660-087-5 (cloth)
ISBN: 978-1-78660-088-2 (pbk.)
ISBN: 978-1-78660-089-9 (electronic)

Contents

Note on the Translation	vii
Acknowledgements	xi
Introduction	xiii
1 Verticalism and Thematism	1
2 From the Molecule to the Organism	29
3 Internal Reproduction	53
4 The Fragmentation and Socialisation of Development	63
5 Signal Stimuli	71
6 Competence	85
7 Autonomous Procedures and Regulated Behaviour	93
8 Open Formations and Markovian Jargon	113
9 'Crossword' Formations	127
10 'Spectacle-Spectator' Complexes	137
11 Forms I, II and III	147
12 The Philosophy of Morphogenesis	157
Notes	177
Bibliography	195
Index	201
About the Author and Translators	209

Note on the Translation

We have no interest in trying to summarise the course of Ruyer's argument in *The Genesis of Living Forms* here, above all since he does such a good job of summarising it himself (in the final chapter) and in carrying the reader along as he, like a detective, tracks down evidence and delivers warrants of arrest. There are, however, a few minor points concerning the translation worth drawing attention to.

Construction. The Ruyerian sentence is not the Proustian sentence, but like Proust, Ruyer makes frequent use of the affordances French provides for long and intricately nested constructions. In the more extreme cases, we have decided that fidelity to his argument supersedes faithfulness to the letter of the text and have accordingly broken these labyrinths down into more familiar English sentences.

Scientific terminology. Wherever possible, we have rendered the many technical terms – from embryology, animal ethology, geology, quantum physics, and the numerous other fields of science to which Ruyer had recourse – with their current English counterpart. Given that he was writing at the end of the 1950s, however, there are some terminological and conceptual anachronisms. In these cases, and wherever the body of scientific research in which he is intervening is likely to confuse even someone broadly familiar with it, we have appended a translators' note, which is prefaced with the initials TN. Translators' notes have also been introduced to clarify some of Ruyer's more obscure, passing references to literary texts and philosophical concepts.

Ruyer's terminology. Most of Ruyer's own terminology is either drawn from the sciences with which he engages or presents no particular problem of translation – the key concept of the book is, after all, that of *form*. As we have just noted, the former cases are dealt with in the translators' notes. With respect to the latter, the following instances are worth remarking.

Ébauche. The English translation of this technical term, which describes the earliest nascent structural formations in the developing embryo, is 'primordium'. While precise, this is lacking the very useful – for Ruyer – connection with 'sketch', the more common French translation. While he only makes an explicit use of this double meaning, it is implicit throughout.

Machines à information. Despite the awkwardness of the term, we have transliterated this as 'information machines'. On the one hand, it does not quite have the generality of the English 'computer', particularly in contemporary technology. On the other hand, the explicit connection between automation and information is important for Ruyer. These reasons, which are touched on at a number of points during this work, are the object of *La cybernétique et l'origine de l'information* (1954).

Power. It is worth noting, finally, that Ruyer almost exclusively uses the French *pouvoir* rather than *puissance* – two terms that are translated by the English 'power'. We have indicated in brackets the single occasion where he uses the latter in this book. All other uses of 'power' should be taken to render *pouvoir*.

Tuyau(x). Ruyer makes use of this term in two general registers. The first has the sense of a manufactured *pipe* or *tube*, such as those which would compose a plumbing system. As Ruyer emphasises in the first chapter, pipes are formed in a factory in accordance with a pre-given mould. The second register is biological, where we have translated it as 'duct'. While this is a somewhat unusual translation of, for example, 'blood vessels' (*vaisseaux sanguins*), it conveys the broad sense of a biological tube. Ruyer makes somewhat humorous use of the same term for these two disparate registers precisely in order to draw attention to the profound difference between them. While a plastic pipe is extruded in a factory according to a pre-established shape, the ducts and vessels of the body arise through a dynamic process of morphogenesis 'guided' by an implicit theme that has no immediate spatial correlate. Indeed, the argument of the book effectively runs from – or through – one *tuyau* and then the next. In any case, the homonymy of Ruyer's usage should be kept in mind whenever tubes, pipes or ducts are at issue.

English terminology. Ruyer makes relatively frequent use of English technical terms throughout *The Genesis of Living Forms*. These are italicised in the original French; we have appended an asterisk to these terms to convey their status – for example, patterns*.

Citations. Ruyer's practice of citation is, to be frank, fairly impressionistic in character, often appearing to be the result of working from memory rather than directly consulting a text. He often conflates different passages, freely paraphrases, and, when translating from English, takes considerable liberties. Our practice has been to give the most accurate version of what Ruyer himself writes rather than attempt to 'correct' his citations, and then, where possible, to append a note giving the details of the transposition. In these cases, we cite from extant English translations. Citations not preceded by this indication are Ruyer's own. We have silently corrected occasional errors in referencing.

In this context, it is worth alerting the reader that a great many of Ruyer's references cite a single volume: *L'instinct dans le comportement des animaux et de l'homme* (Paris: Masson et Cie, 1956). This text collects the proceedings from a 1954 conference on animal behaviour that took place at the Fondation Singer-Polignac in Paris, at which Ruyer also presented a paper titled 'Finality and Instinct'. The other twenty-one contributions in this text range across physiological, psychological and more broadly philosophical terrain, and Ruyer makes reference to almost every main theme raised in them. It would not be a stretch, therefore, to characterise *The Genesis of Living Forms* either as the result of the provocation of this conference or as an extended critical appraisal of the state of the field represented by these expert contributions.

Examples. Ruyer's texts occasionally include short sections presented in a smaller font size. These parenthetical moments are for the most part an opportunity for Ruyer to consider a particular example or set of examples that support his argument in the main body of the text.

Pagination. The marginal paginations included here refer to the original French version of *La genèse des formes vivantes* (Paris: Flammarion, 1958).

—Jon Roffe and Nicholas B. de Weydenthal

Acknowledgements

A version of chapter 12 was published in *Parrhesia* 29 (2018). We would like to thank the journal and the editorial board for permission to republish it with changes here.

The translators would like to thank Rowman & Littlefield International, particularly the long-suffering Rebecca Anastasi and Sarah Campbell. We would also like to thank the rest of the editorial board of the Groundworks series – Arne De Boever, Bill Ross and Ashley Woodward – for their patience and support. Anne Sauvagnargues's enthusiasm for the book was a welcome motivation at a difficult stage in its production.

Jon would like to thank David Rowe and Christopher Pollard for their timely help and Isabelle McGovern for her companionship during much of the translation's production. Nicholas would like to thank Alice Grinton for being so patient and Nina B. de Weydenthal for being the book's guiding theme. This is a book for her and about her.

Introduction

Morphology, the study of forms and their arrangement, presents no fundamental difficulty. It requires more than just precision or meticulousness. Quite often it requires the kind of ingenious, indirect methods which have led to the structural schemas of organic chemistry or to the cartography of genes in cellular nuclei. The results of these indirect methods, however, are often then directly verified. Photographs of crystalline lattices taken by electron microscopes sometimes reveal to us structures that were formerly the subject of ingenious conjecture. This demonstrates that, at least in principle, morphology is 'easy' – in the very specific sense given to the word in scientific research – easy as a vision, easy as a direct description.

On the other hand, the science of forms allows us to escape from the disagreeable obligation to engage in philosophical subtleties concerning the value or even possibility of knowledge [*connaissance*]. In his last work, Eddington describes the intellectual event that the encounter with Bertrand Russell's theory of the structural character of scientific knowledge represented to him.[1] Most philosophical speculation could be avoided, Russell says, if the importance of structure and the difficulty of going beyond it has been recognised.

> For example, it is often said that space and time are subjective, but they have objective counterparts; or that phenomena are subjective, but are caused by things in themselves, which must have differences *inter se* corresponding with the differences in the phenomena to which they give rise. Where such hypotheses are made, it is generally supposed that we can know very little about the objective counterparts. In actual fact, however, if the hypotheses as stated were correct, the objective counterparts would form a world having the same structure as the phenomenal world [. . .] In short, every proposition having a communicable significance must be true of both worlds or of neither.[2]

In other words, while the dog we see is not the dog as 'animal-in-itself', both dogs have four paws, a tail, lack sweat glands, and share all of the other anatomical details arranged in the same order. It matters little that our world is a world of shadows if, as in Scarron's hell, the dog's shadow trots behind the shadow of his master on four paws like the real dog.[3] It matters little that

I only know of the real and living brain of the dog in terms of the perception of this brain in my own so long as I am capable of describing its anatomy and functioning with precision. Many philosophers, faced with 'the problem of two dogs', have the impression that the scientific point of view must be rejected as impossibly naïve and that it is necessary to think of the notion of phenomena in more direct and subtle terms – for example, by returning to the immediately given and devaluing as artificial everything that science has accomplished in the deciphering of sensory experience.

But this discouragement – or pretension – is completely unjustified. The theory and practice of information machines have familiarised us with the decisive importance of structural correspondences. The movie-goer is indifferent to the manner in which the soundtrack of a film is obtained, and to the materials used to obtain it, so long as the sound is faithfully reproduced. The listener is indifferent to whether a concert played on the radio is transmitted by amplitude (AM) or frequency (FM) modulation, so long as it sounds good. In the same way, the anatomist is indifferent to whether the structure of the dog is known only through its cerebral relays or directly as an absolute phenomenon. And as long as the structural group can be 'decoded' in the end, he will be equally indifferent to the discovery that we are in Plato's cave, or in Kant's world – or Berkeley's, or Husserl's.

Naturally, everything that can be said about the knowledge of structure can equally be applied to the knowledge of functioning since the two come to the same thing. In mathematics as in biology, a structure is a closed group of possible operations. The rotation of a sphere, as a group of operations on the points of a sphere, precisely defines the structure of the sphere. The dog's mode of locomotion comes to the same thing as the structure of its limbs, or more precisely, to the structure of its limbs *plus* the structure of the nervous apparatus that commands the muscles. In any case, scientific physiology postulates that the means to completely account for the dog's mode of locomotion as a cyclical functioning can in principle be found within the actual structure of the nervous system. This postulate can thus be expressed in the following form: 'For any given functioning, one must always be able to conceive or create an automaton capable of equivalent structure and functioning'. The automaton will of course be made of metal or plastic rather than of living cells, but its structure, according to this definition, will be exactly the same, with respect to the relevant functioning, as the structure of the living dog.

If morphology, along with functional physiology, is the simplest part of classical science, morphogenesis presents, on the contrary, the greatest difficulty and even mystery, and it is easy to see why. If knowledge [*connaissance*] rests on structural correspondence – on the isomorphism between real object and its phenomenon or theoretical schema – how could it be possible to speak of any isomorphism between structural schemata at all, or the passage from the *absence* of such a structure to its *presence*? This dog, which we know once existed in the form of a unicellular fertilised egg before its four paws and its nervous system formed, cannot be understood in the same way that we understand how the dog, which *now* has four paws and a nervous system, is capable of walking. *There can be no isomorphism between a form and a formation, but only between form and form, or formation and formation.*

Faced with the mystery of morphogenesis, there are only two possible approaches: to attempt to deny formation by reducing it to a functioning, or to appeal to a non-structural schema, to an analogy with another, more familiar domain, in which formations are also observed, such as the domains of artistic or technical invention. According to the latter hypothesis, the structure and functioning of the automaton correspond to the anatomy and physiology of the dog, and the formation of the dog corresponds to the invention of the automaton. The isomorphism of knowledge [*connaissance*] is preserved: in formation as in invention, there is a passage from the *absence* to the *presence* of structure; or, alternatively, there is a passage from an iso-amorphism to an iso-morphism. But this is at the price of renouncing scientific *knowledge* [*connaissance*] of both invention and formation. Scientifically inclined psychologists have not lost hope of explaining invention as a function of the human brain. It is clear, however, that this hope must be renounced if formation and invention are to be conceived as analogues, for it is not the brain, human or otherwise, in which the formation of the dog, brain included, originates. Nature does not have a nervous system from which it forms nervous systems. As Plotinus said, it needs no hands with which to make hands.[4]

Chapter 1

Verticalism and Thematism

FUNCTIONING AND FORMATION

The fundamental feature of organic formations can be metaphorically characterised as a 'verticalism'. Irreducible to function, formation can perhaps be said to be 'perpendicular' to it. A glance at the diagrams (figures 1.1–1.7) gets to the point more quickly than the explication that follows. Regardless of whether we consider the organic formation of a circulatory canal for air or blood, or an object of human creation, what is important is the opposition between the diagram's vertical and horizontal arrangement. The first represents a formation, the appearance of new forms that nothing – except for the analogical knowledge [*connaissance*] of similar phenomenon – would allow us to deduce from the initial givens. The second represents the functioning of structures after their formation in space and time, a functioning which can easily be deduced by considering these structures. Every treatise on embryology includes a profusion of examples of vertical diagrams. Every treatise on physiology includes examples of horizontal diagrams.

In both cases, a given state orders the one that follows. *The notion of the linking of forms is more general than that of functioning*; there may be a linking together of forms without any functioning. There is never any pure emergence or appearance, in the sense in which we say that a ghost appears, or in which Venus appears above the stream 'like a mist'. A gutter is formed *from* a flat panel, and the pipe is formed *from* the gutter.

PSEUDO-FORMATIONS

It is certainly the case that nothing but functioning takes place in a factory that manufactures gutters or pipes by stamping them out. On the other hand, when ocean sedimentation, viewed laterally, flows onto a continental shelf, forming folds which sometimes closely take on the appearance of a gutter – in short, what could be called the 'morphogenesis' of a mountain range – nobody would see anything but a mechanical function, and certainly nobody would have recourse to a mythology of a divine, 'vertical' and 'artistic' creation of mountains in anything but a religious or poetic sense. But organic morphogenesis, even when it is very close to functioning, is something completely different because it results not only in the transformation of an initial form, and not only a brute increase in complexity that would be perfectly accounted for as a 'quantity of information', but in an increase in complexity in a self-sustaining, consistent, unified totality, capable of serving as the basis for a new formation in its turn. The brute increase in complexity in an open ensemble is not a sufficient criterion: it would probably take around the same number of words to telegraph a description of the Alps as it would to telegraph a description of mammalian embryogenesis. But once formed, the mountainous folds can only then function, that is, be passively subjected to pressure, erosion and chemical decomposition while the organism continues to be differentiated. Furthermore, at the moment of the formation of a mountain's folds, we encounter the play of enormous exterior lateral forces on the sedimentary bed that press forward step by step. In the folding, invagination and migration of organic tissues, on the contrary, the forces in play must first be created or mobilised on the spot through a first differentiation. They are the 'effects of explication' as well as 'explicative causes', representing not

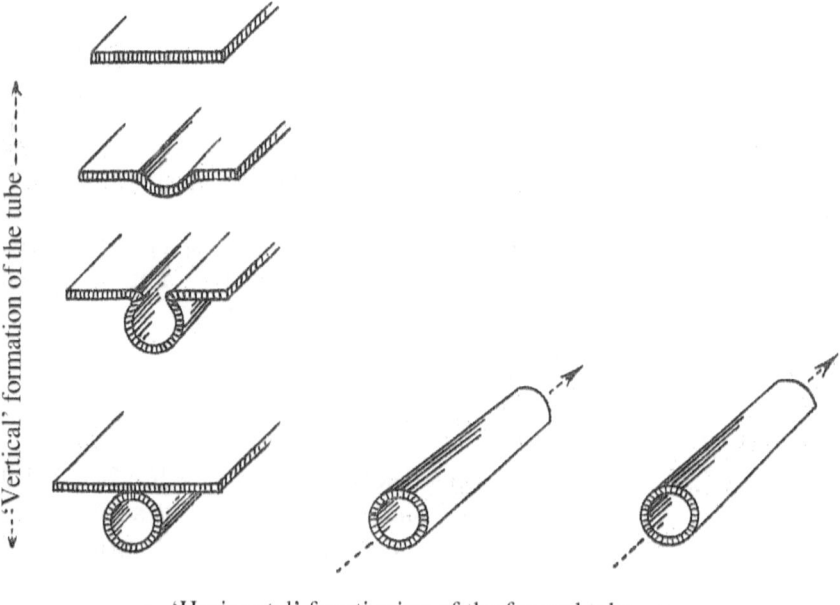

Fig. 1.1

an enormous dynamism but, to use Dalcq's expression,[1] an organisational dynamic – 'organisational' by virtue of the fact that it is already organised, distributed, hierarchised and synchronised.

There is an analogy between organic formation, as opposed to functioning, and the assembly of a new factory or production process involving new machinery, as opposed to the functioning of an established factory or processes of fabrication under way. Accelerated functioning increases *production*, but not the *productivity* of either the factory or the organism. Once established, the factory which fabricates gutters by stamping them out 'functions', but the assembly of a production process cannot be a functioning. It must have been an organisation, a true formation.

FORMATION OF THE AMPHIOXUS

Let's recall here in broad strokes the early moments in the development of the amphioxus, which presents us with a kind of schema of development for all vertebrates. Without growing, the egg is segmented into two, then four, then eight cells of roughly equal size. Those cells that continue to divide constitute a small sphere the size of a blackberry (*morula*) before becoming a hollow sac (*blastula*).

The lower section flattens, collapses and subsides (*gastrulation*) into the higher hemisphere as if an invisible thumb were pressing on a rubber ball (figure 1.2). The cavity of the blastula is then reduced and a new cavity is formed, one which constitutes the primitive intestine. This primitive intestine communicates with the outside through the residual orifice: the blastopore. The area of the blastopore, above all on the dorsal side, is in fact the active agent of gastrulation. In the gastrula, the primordia of the animal have already been put in place. The gastrula has a cephalic-caudal axis with a bilateral symmetry and a dorsal-ventral axis. The parts which are invaginated, however, do not share the same destiny. The dorsal arch constitutes the chordal process,[2] and in the mesoblast, the ventral side constitutes the endoblast

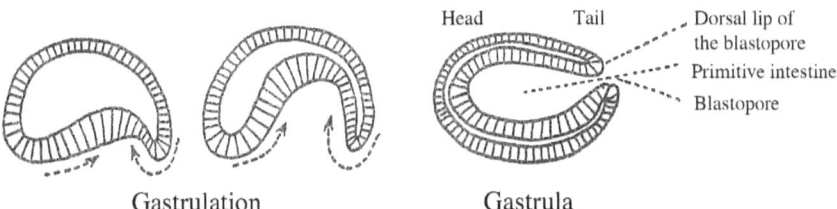

Gastrulation   Gastrula

Fig. 1.2

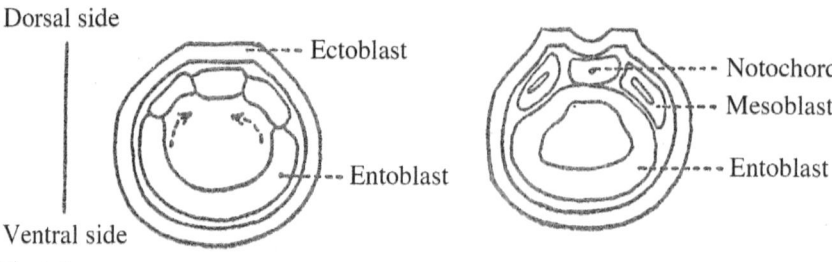

Fig. 1.3

which soon detaches itself from the mesoblastic primordia and is fused beneath the notochord (figure 1.3).

On the dorsal side of the embryo, in contact with the notochord, the nervous system begins with two crests which appear along the length of the cephalic-caudal axis, bringing together and constituting, by fusing together, the neural tube. It must be noted that if, using the method of fate mapping,[3] the cells which, once invaginated, will become the notochord, the mesoblast, or the intestine, can already be identified, this does not mean that they contain within themselves the structures in a preformed state for if these cells are experimentally displaced, they can yield something other than what they would have done had they been left in place and the embryo had developed normally.

The development of the amphibian egg is closely analogous but for the fact that being richer in reserved nutrient matter, segmentation and then gastrulation are somewhat impeded by this mass of vitellus and the *dorsal* side of the blastopore is rather more active (figure 1.4).

Even in this rough-and-ready schematisation, it is difficult to overlook the characteristic nature of creative work, free behaviour, formation irreducible to functioning, or, in short, 'verticalism'. In gastrulation, all is activity. If the blastula were a system in physical equilibrium, it is hard to see how its manner of pursuing the path that leads to such complication could be understood.

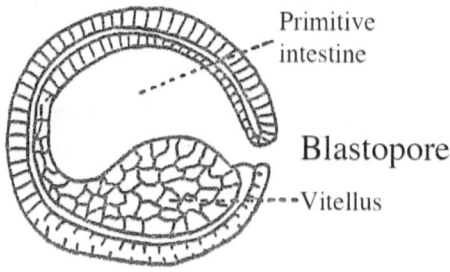

Fig. 1.4

GRAFT AND PROSTHESIS

The progression and the transmission of forces or substances in horizontal functioning can be followed, allowing for the intervention by human technique and by way of more and more daring partial substitutions. We can already temporarily – and, soon no doubt, permanently – replace a failing organic canal with one made of plastic. In vertical formation, we can also discover chains of effects and the progression and transmission of substances in which intervention is possible. Experimental embryology presents us with techniques at least as daring as the techniques of prosthesis. In the precocious stages of a differentiation, for example, we can surgically transplant a graft from one part of the developing organism to another part, onto another organism of the same species or even onto an organism of a neighbouring species.

In a great many of these cases, normal development will nevertheless continue, the rootstock imposing on the graft its own mode of differentiation – here, for instance, contributing to the formation of a gutter (figure 1.5). Even in cases in which the graft does not completely conform to the inductions of the donor, developing as it would have in its original place, the two adjoined tissues at least partially harmonise with each other.

But it is impossible to confuse these two kinds of technique, vertical and horizontal, graft and prosthesis. In a 'vertical' intervention, the surgeon counts on a competence possessed by the tissues involved. Through the displacement, he modifies the possibilities for the exercise of these competencies. He is the equivalent of a CEO who can modify the situation of his workers by allowing them to use their competencies to a greater or lesser extent; or, again, by modifying the signals that regulate their work, or modifying the rhythm according to which the materials they work on will arrive. He is not the equivalent of an engineer imagining machines that could substitute for the – already automatic – work of manual labourers or unqualified workers.

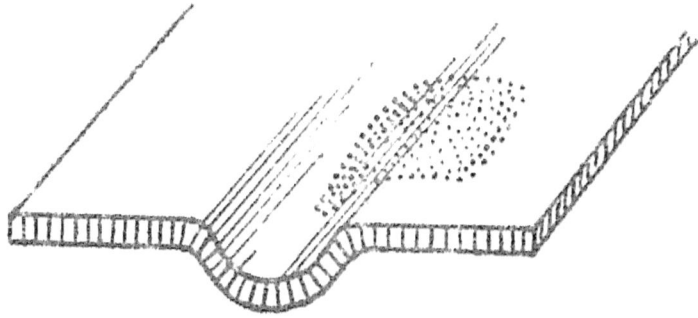

Fig. 1.5

FUNCTIONING, MORPHOGENESIS AND BEHAVIOUR

Behaviour can be understood as a synthesis of functioning and formation. A machine, left to its own devices, functions. But when it is related to a theme or intended for a use, it has a 'role' that it 'performs'. Behaviour, contrary to the thesis of mechanist behaviourists, is irreducible to a functioning. It implies improvisation and inventive adaptation to a function-role. In the hands of a pilot, an automobile, boat or plane behaves well or poorly depending on the pilot, whose intention thus takes the place of a vertical theme in a formation. A living organism, acting instinctively, also 'behaves'. This is what signifies that a formative and improvisatory component which breaks with the functioning of the organs is already present: the 'pilot' is one with the machine. But at the same time, the pilot is an engineer whose improvised behaviour not only uses the machine but guides it, corrects it, completes and perfects it. Behaviour does not therefore depend exclusively on a structure – it is also the improvisation of structure. There is formation in all behaviour, and behaviour is the principle of formation.

In the higher animals, behaviour is related to the nervous system, which became the specialised organ of diffuse auto-conduction in the embryo. When a higher animal exhibits behaviour of some kind, its nervous system alone manifests the formative improvisatory component that is present in all behaviour while the rest of its body is constrained to function according to the guidance of the nervous system. In the behaviour of a protozoan, the entire organism at once functions and improvises. The unicellular organism improvises its pseudopods in order to move around, and its mouth and digestive tube in order to eat. In the embryo, too, even that of a higher animal, improvisation is everywhere implicated in the functioning of the organs already acquired through preceding developments. The blastula improvises – according to a specific theme, certainly, but beyond any automatic functioning – like the cell migration which transforms it into the gastrula, like the unicellular organism improvises its pseudopods. Behaviour is indiscernible from development, of which it is at once the principle and the manifestation. In the lower animals, bacterial colonies or amoebae, development remains indiscernible from behaviour until the death of the organism.

In an adult endowed not only with a nervous system but a brain with a developed cortex, as in a human being, it is extremely common for the whole of the organism to be represented multiple times by projection onto the cortex, like a sort of *homunculus*: sensibility is projected onto the parietal cortex; motricity, onto the frontal cortex; the visible body onto the occipital cortex; and even, in a still less precise fashion, sensibility and emotive motricity onto the pre-frontal cortex.[4] In psycho-social behaviour, the functioning of the body is subordinated to the immediate behaviour of the 'projected body', of the cortical *homunculus* that behaves in the manner of a unicellular organism or an amoebic colony.

Consider a human being walking towards a determined goal on a poorly marked path requiring a certain inventive vigilance. The muscles of his limbs, insofar as they obey the impulsions of the nervous system, are equivalent to functioning. But the cerebral, sensori-motor *homunculus*, the zone from which nervous impulsions depart, cannot be limited to functioning: the cerebral *homunculus*, the ensemble of neurons woven into a network that constitutes it, forms a sort of enormous cerebral amoeba[5] which 'behaves' according to an immediate auto-conduction analogous to the amoeba emitting pseudopods and completely transforming its apparatus as it moves. The voluntary locomotion of a human being is thus the functional amplification, obtained through complicated relays, of an improvised schema of movement in the *homunculus* of the motor cortex. In an adult multicellular organism, the brain is an apparatus which permits the separation of the 'functioning' component from the 'behaviour' component, which in the unicellular organism or the embryo merge together. The brain is restricted to functioning no more than the embryo is. This is why, for example, if a faradic current[6] is applied to the motor zone of a living brain, variable motor 'responses' are produced rather than stereotypical movements. The cerebral *homunculus* is only localisable precisely because it is behaving or in perpetual development, like an amoeba or a colony of amoebae in a crystalliser.[7]

Without succumbing to a vicious circle or a contradiction, we can assert that there is no difference between the claim that behaviour is a synthesis of formation and functioning (above all if we think of behaviour arising from a nervous system) and the claim that behaviour is the principle of functioning (above all if we think of beings that are not endowed, or not yet endowed, with a nervous system).

Should one wish to study the embryology of behaviour in the psychological sense of the word, as A. Gesell does, it becomes apparent that it is impossible to dissociate behaviour and morphogenesis.[8] The thesis according to which biological heredity creates the organic structures which *subsequently* determine behaviour is today recognised as false. Heredity is at once a factor of structures and behaviours, or, to be more precise, formative behaviours and instinctive behaviours. More than a parallelism, there is an identity in nature between modes of development of the structures of the body and the modes of development of instinctive behaviours. 'Vertical' development is a behaviour, and instinctive behaviour develops in the fashion of an organic draft or primordium. The two are strictly interwoven, often through the convergence of fragmentary developments. In the ontogenesis of young passerines, the behaviour 'scratching the head while balancing on a foot and a wing' first appears in fragments. The parts of this complex movement appear before the behaviour, as a functional unity, can be accomplished. Ten-day-old marsupial foetuses (the opossum in particular has been studied by Hartman and McCrady), which resemble earthworms more than mammals, are capable of crawling from the urogenital canal to the pocket by using their anterior membranes and moving their heads from side to side until they arrive at the teat, where they attach themselves as if to an umbilical cord in order to complete their development. This behaviour makes use of several functions but envelops and inserts them in the developmental process as one phase amongst others. The same applies, as recent microfilm techniques and applications have shown, in human embryology. The behaviour of the tonic neck reflex, implying dissymmetrical movements of the shoulders and arms, and which will become the constitutive element of a group of subsequent behaviours, is already sketched out at eight and one-half weeks when the human foetus is no longer than 25 millimetres in length.[9] The freely floating embryo itself probably contributes to its equilibrium, given the very early growth of the semi-circular canals, but without yet being any more than a primordium.

Structure and function, structure and behaviour, far from being equivalent to 'machine' and 'functioning' (functioning only ever being defined *in terms of* the structure of the machine), develop in concert, often in the same step, function and behaviour always anticipating structure *a little*. Function does not *greatly* anticipate structure – for we would then depart from the continuity of 'verticalism' and fall for a kind of magical or miraculous appearance. But in order for there to be any development or behaviour, function must anticipate 'possible functioning' *a little* – the structural base would not otherwise have changed. Development cannot, like functioning, be cyclical, always returning in principle to its starting point. If, at the end of embryonic development, new structures, obvious and widespread, are there, the infinitesimals of novelty would have had to be integrated. One can only live on credit.

FORMATION AND WEAR

In a machine, functioning is not rigorously cyclical since the machine wears out and finally breaks down. But as far as we know, nobody has yet been so bold as to straightforwardly maintain – though it has often been done in a surreptitious fashion – the thesis that development is only a phenomenon of wear and that the adult organism is a worn-out egg or an embryo. An old man is perhaps, to a certain degree, a worn-out adult in the mechanical sense of the term, and it is quite likely and even probable that wear through aging begins very early in all constituted and mechanised structures. But morphogenesis or regeneration are precisely opposed to this wear, which is first dominated, and then equal, before being predominant. The opposition between the two factors is self-evident, and there should be no question of confusing them. If, consequently, the non-closure of the cycle in behaviour and development is not explained by wear, it must rather be explained on the basis of an infinitesimal anticipation of formation in advance of functioning, of a sort of *negative wear*, or again, when wear is a phenomenon of entropy, of a 'negentropy', or an *input* of form. The old vitalist definition, 'Life is the totality of forces which resist death', is at once incontestable and insufficient. It is only true with respect to the ideal moment in which vertical development, through its morphogenetic contribution, compensates for exactly this degradation of structure produced by horizontal functioning. But this ideal moment does not exist, at least for the organism grasped as a whole, since the powers of regeneration are always very unevenly distributed. And above all, an organism which constructs itself does not limit itself to resistance – it creates forms.

ACTIVE DEPLOYMENT

That function anticipates structure is borne out by the facts. Function and behaviour equally anticipate structure such that the final differentiation often appears to be the active deployment of more refined means, the perfecting of a crude and already active function. As in a human factory, 'automatisation' in the organic 'factory' is often a belated development.

> A primitive organic rhythm often precedes the active deployment of nervous mechanisms which will come to control them and which will automate, in the mechanist sense of the word, the autonomous agents that they were. In the adult organism, the beating of the heart seems to be causally explained by the purely physical action of nervous influx. But how can the fact that the heart of a chicken embryo, on the fourth day of incubation, already presents an electrocardiogram identical to that of an adult on the basis of its own muscular cells and thus without the control of a nervous system be explained?[10] The active deployment of means does not itself happen without its own means, which appear as causes to those who are absolutely determined to consider them as such. But it is striking that active deployment already has the appearance of a formative behaviour. Nerve cells, which will establish the auxiliary mechanisms of behaviour already dynamically sketched out [*ébauché*],[11] themselves behave just as instinctively. Sped-up microfilm shows their growth cones searching, with amoeba-like movements, for a path through the intercellular spaces of living tissue, guided by the signals-stimuli emanating from the muscle cells to be innervated.[12]

THE MATURATION IMPLICIT IN FUNCTIONING

The schema of verticalism must not lead us to believe that the constitution of an organ, and then its functioning, always takes place in neat, successive and distinct phases. Often there is even a certain incompatibility between the work of formation and the work of functioning. In embryogenesis, the pulmonary artery, the lungs and many other organs are formed before they function. But even more frequently, functioning is closely implicated in formation: the umbilical artery and the heart are formed and function at the same time, as is the case for the greater number of organs which function from the first phase of their formation and which continue to be formed while functioning.

An adult hand functions and is used as if it were an undeveloped organ (even though it imperceptibly repairs itself, for example, in the growth of fingernails and more generally in the incessant flux of molecules which circulate throughout its form). An embryonic hand is developed without being used. But the hand, eye or nervous system of a newborn is fully developed and ready to use. Functioning does not wait on the complete constitution of an organ. A general maturation-development continues from the first use. Furthermore, this first exercise is often indispensable to the good continuation of development. A young animal that is prevented from seeing, its eyes covered by a screen, will suffer incurable ocular damage.

THEMATIC CONTINUITY

In the two directions of the schema of formation, vertical and horizontal, whether in a phase of morphogenesis or in functioning, a passage in time from one structure to another appears to be involved.

In morphogenetic terms, though, there is a *thematic* continuity, while in terms of functioning, there is a *positional* continuity. If it were not so difficult to impose an absolute terminological rigour on this point, the words 'form' and 'structure' could be made more precise by noting that vertical morphogenesis manifests a thematic continuity of forms and that horizontal physiology manifests a permanence of the structure despite changes of position. *In the first case, there is a passage from one form to another; in the second, there is a passage from one set of positions to another.*

In the course of a development, there is always a passage from the primordium of a completed organ (for example, from the primordium of a tube to a completed tube). More precisely, there is a passage from a so-called *presumptive* area, which the biologist knows by analogy will normally produce a defined organ in the same so-called *determined* area, about which we not only possess subjective presuppositions concerning its terminus of development but are the actual recipients of (or are connected to) a theme of differentiation, as subsequent events demonstrate. At this stage of determination, differentiation is not yet observable, but the objective character of the determination can be demonstrated by experiments in transplantation: if the graft develops *according to its place of origin* and not *according to the place to which* it has been transplanted, it is there that it is determined. For example, if, in amphibians short of the well-defined state of gastrulation, a tissue sample which would normally produce a neural tube is grafted into another embryo in the presumptive region of the formation of gills, it will produce gills. The presumptive epidermis, transplanted to the area of the presumptive neural tube of a second embryo, will develop into a spinal cord or part of the brain. But a little later, after this 'something' that is called determination, this miraculous plasticity is lost: grafts of the same kind no longer produce the same result; grafts develop in their new place as if they had been developed *in situ*, without adapting, and according to their initial theme, before the intervention of the experimenter.

THE STAGES OF DETERMINATION

This restriction of morphogenetic capacities to a tissue, and the correlation of the progress of determination and differentiation, operates in stages. Thus the ectoderm of the gastrula, which itself results from a determination of the blastula, is first determined in the direction of epidermal, neural or mesodermal differentiation. Each of these broad categories is then divided into secondary determinations: for example, the epidermal determination leads to the true epidermis, or to the primordium of a lens, or a tympan, etc.

In the same way, a limb bud is first determined as a leg, then as a right leg, etc. Waddington, Needham, Lotka and others have represented this succession of determinations as a series of passages through increasingly stable states of equilibrium. Take three layers, or truncated cones (figure 1.6).

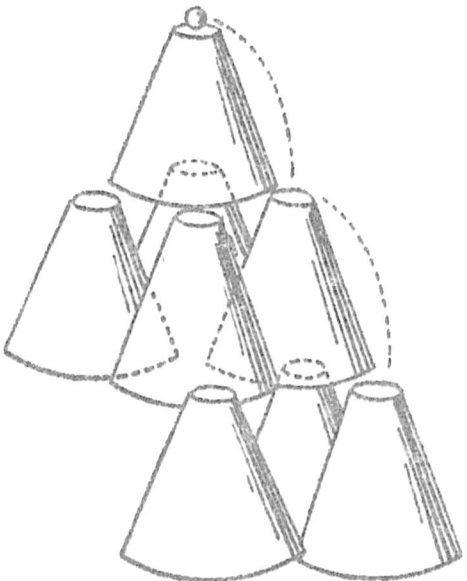

Fig. 1.6

A marble is first placed at the summit of the highest cone, from which it can fall, subject to the momentum it receives, onto one of the cones of the next level; then, continuing to fall in the same way, it attains a plane of absolute stability, which is the adult organism. This illustration has the value of representing the irreversibility of differentiation after determination. As we will see, it also allows us to introduce the definition of successive 'flicks of the finger' which will make the marble fall in this or that direction from the summit. But it also risks misleading us in that it characterises determination, qua restriction of 'power', in strictly negative terms. A given bud *could* become a left or right leg; after determination, it can no longer produce a right leg. But the words 'power' and 'potentiality' have a double meaning. They designate either a non-dynamic possibility (apprehended by a witness) or a 'dynamic power'. When the transplantation of a graft shows that the 'presumptive' fate of a given tissue sample is modifiable and that it *could* therefore become something other than what it would normally have become, nothing permits us to take 'power' in the dynamic sense and to say that the tissue was capable in itself of differentiating in multiple directions but only that determination *limited* this power. This presupposition, as we have stressed, belongs only to the perspective of the biologist. Determination, on the contrary, implies a positive and dynamic power belonging to the tissue itself, which is immediately manifested in its differentiation, and which actively conforms to a formative theme. The schema of the marble and the cones combines these two senses of 'power', assimilating them to the 'potential' energy of a physical body standing at a certain height and progressively losing this energy to the degree that it has descended.

And yet, demonstrably, the positive facet of determination appears through the creation of a *more complex* rather than a more primitive form. To present a theme to an artist would perhaps, on the one hand, prevent them from working on another theme that they *would have been able to* produce for themselves. On the other hand, however, it is certainly the case that it provides them with something that actually guides their creative effort. The biologist's spontaneous use of the word 'sketch',[13] borrowed from art, to designate the primitive stages of a differentiation shows that faced with the facts, they do not hold to the image of a marble which experiences a *loss* in elevation but instead often have recourse to the image of a drawing which *gains* in detail. An embryo sketches an abstract and thematic 'assembly': an axis of bilateral, cephalo-caudal or dorso-ventral symmetry. The nervous system, destined to become more complicated than a telephone switchboard, first resembles a simple gutter and then a tube. The human hand first resembles a paddle of which the buds of fingers are all similar. The fall of the marble is a degradation of energy, a diminution of information and an augmentation of entropy, or disorder, while the succession of determinations is an augmentation of information and a creation of structural order.

DETERMINATION AND DETERMINISM

The marble diagram also risks giving the completely misleading appearance of mechanical determinism to determination. Development certainly no longer appears to possess 'freedom' – it follows precise rules. Embryology is a science, and the biologist can predict in general terms the behaviour of an embryonic area or a graft. But neither can it any longer appear to conform to the outline of a physical system submitted to determinism. What gives it such immense interest is precisely that *it can be recuperated by neither the category of 'freedom' nor the category of 'determinism'*. Freedom exists (theoretically) when a being, x, in a given situation, invents or improvises a suitable behaviour that is not rigorously deducible from its situation. Determinism exists (theoretically) when, from the positions and movements of a group of particles, their subsequent positions and movements can be rigorously deduced. The marble example does not seem to be determinist since we cannot deduce the direction of the marble's fall from its position at the summit of the cone. But in the minds of orthodox biologists, the orienting flick of the finger is provided by a chemical inductor, and the action of the inductor, *plus* the reaction of the tissue, is supposed to completely explain morphogenesis. We are therefore indeed within a determinist schema: the current situation (here, the chemical state of the inductor *plus* the chemical state of the tissue) determines, without residue, the subsequent situation. But however quick embryologists are to theoretically postulate a crypto-determinism, in their real practice they never in fact consider the situation – that is, the state of the organism at a given moment – in the way that an astronomer considers the situation of the planets in order to calculate their subsequent positions. Neither do they ever consider in general terms the physico-chemical situation in the way an engineer does, presuming the known result of an equation on statistical grounds. Finally, they never consider the situation to be that of a structured machine or a field of forces whose functioning or effects could be calculated. The embryologist speaks of a coordinated system of primordia; he predicts their subsequent fate by analogy with organisms of the same species, and if he intervenes in this development, it is not at all in the same way as a chemist introducing a new body or physical variable into the course of a reaction. The facilitation of the marble's fall from one stage to the next could also appear as the action of a catalyst. This represents the action of a biological inductor very poorly and the subsequent differentiation which goes from form to form, from form in outline to achieved form, even more poorly again.

THEMATIC FORMS AND DEFORMABLE SYSTEMS

Neither development nor biological phenomena in general can be understood so long as we do not reject the postulate, borrowed in its entirety from classical physics, according to which a form is only a set of positions coordinated according to step-by-step relations. Even those biologists aware of the insufficiency of classical physics in biology have not always clearly seen where the error lies. It is easy to imagine 'deformable' mechanical models, that is, models in which the same form or its analogue is maintained throughout a series of modifications provoked by a local action as if a formal theme dominated the possible positions. The pantograph is an example.[14]

Let us first consider a biological example, borrowed from H. J. Jordan, that appears analogous to the pantograph.[15] This is the example of the determined forms that the leaves of the *Sagittaria* adopt in response to a simple cause. Submerged, the plant's leaves possess a thin, lance-like appearance; when exposed, they take the form of an arrowhead. Causal analysis shows that the principal cause of this difference is light. When it is weak, it gives rise to lance-like leaves, and when strong, the leaves take the form of arrowheads. If the plant's organs were independent, such a simple cause could only destroy the given static order, that is, give rise to chaos. 'From the leaf, which possesses a given order, another order always arises'.[16] Is this case the same as that of the pantograph? Clearly not. In the leaf and in the organ in general, and in both local and global reactions, development passes 'from form to form' and not 'from position to position' and is due not to mechanical connections but rather 'to the initiation of certain complex and harmonious movements'.[17]

THEMATIC FORMS AND FIELDS OF REGULATION

But many theoreticians, notably those under the influence of Gestalt theory, have thought that by substituting a mode of dynamic liaison (for example, of equilibrating forces within a field) for a mode of mechanical connection, they have improved these models in a decisive way and have understood thematism.

Take for example one of the most extraordinary discoveries concerning development: the *splitting* of a primordium, or the *fusion* of two primordia. We know since Driesch that if one of the first two, or even first four, cells of a sea urchin egg are isolated, this half or quarter, normally destined to produce a half or quarter of an embryo, regulates and produces an entire well-formed, if somewhat smaller, embryo. Driesch describes his stupefaction in the face of this discovery:

> I note that, on the night of the first day of the experiment, the hemispheric half-embryo regained the curvature of a complete sphere of a smaller size, and the following morning, there was a complete blastula. But I had been so biased . . . that I was still waiting, on the following day, to discover half an organisation, half an intestinal tube, half a mesodermal ring. But the gastrula would develop into a complete animal, and it would produce a larvae that was small, but complete and typical.[18]

Conversely, if two whole eggs are adjoined, the ensemble produces a unique embryo, one simply larger than the normal embryo. And furthermore, if – in subsequent development – an accident or an experiment eliminates the intermediary tissue between the primordia of a pair of organs and if the two thus come into contact early, only one organ will be produced in place of the normal two.

The model of the pantograph or Jordan's deformable triangle cannot explain facts of this kind. But if we consider a physical system characterised by dynamic bonds – for example, an electrical field in a capacitor, a magnetic field, a soap bubble or a liquid crystal – facts analogous to fusion or splitting, in short, the conservation of a formal theme, are clearly apparent. In baptising all thematic regulation as field phenomena*, embryologists allude to physical fields of this kind, without seeing that there is no significant relationship between the simple, regular distribution of substances or forces and the possibility of maintaining a complex form in every part of an organic field. Biologists, as Waddington says, use the notion of 'field' like a sort of 'wild card' which allows them to explain almost anything.[19] The notion has thus been, for them, the means of introducing thematism but without properly understanding and by imagining that they have adopted it in terms of a physical model.

LIQUID CRYSTALS

Overly simplistic models of the 'soap bubble' or 'magnetic field' kind can be set to the side, but liquid crystals, in which molecules are oriented in a single direction and not in three as in ordinary crystals, are of more interest since, as we have seen, they can be more than just a 'model'. They certainly exist in living cells.[20]

Now liquid crystals, for instance ammonium oleate, can, if they are broken up, regenerate into two smaller crystals or, if they are joined together, fuse

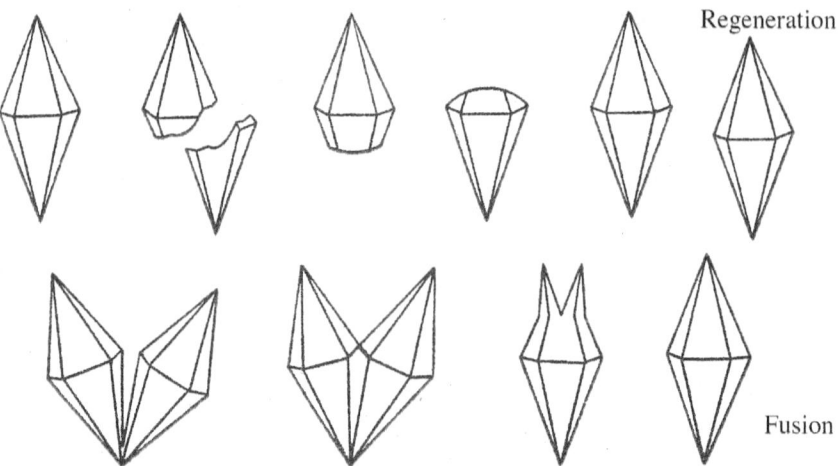

Fig. 1.7

and produce a single large crystal. The parallel with Driesch's experiments is striking. Driesch was stupefied by his experiments to the point of using the familiar phrase 'I just can't get over it'.[21] The equivalent of a man converted by witnessing a miracle, he passed from positive science to if not religious faith, then at least philosophical faith in a transcendent principle. If the models of liquid crystals prove to be illuminating, though, his 'stupour' will have been a poor guide.

Now, there are good arguments in favour of this view. Many organic fusions apparently depend on orienting factors of a physical kind. G. Teissier has fused together larvae of hydra, and the success of the fusion depends on the alignment of the antero-posterior axes. The unstable bi- or multi-composite monsters that Fauré-Frémiet obtains with the ciliate *Urostyla* strangely resemble crystals of ammonium oleate in the process of fusion.[22] But it is the work of R. G. Harrison which is of primary significance. He attempted to interpret determination itself according to the crystalline model, and if his conclusions had been true, he would have shown that it is not thematic, or at least that it would only be thematic in the manner of a physical order. The determination of a limb, as we have seen, takes place in stages. By transplanting, and intervening in, the various stages of the axolotl's development, the discs representing future limbs, we perceive that the determination of the dorso-ventral axis is earlier than that of the medio-lateral axis. At a certain moment, for example, if the antero-posterior axis is inverted, being rotated around the disc, it produces an arm with an elbow pointing outwards rather than inwards, while if the other axes are inverted by transposing the left side to the right while maintaining their anterio-posterior direction, the limb develops normally. The auditory vesicle which will give rise to the ear develops in an analogous fashion: the sketch is first indifferent to directionality. Later, the three axes are successively determined in the same order as they are in the limbs. Other biologists have shown that the same holds for the tail, for the early kidney and even for the neural primordium, in which the anterio-posterior axis is fixed before the others. That such different systems develop in the same order seems to indicate that this order depends on simple physical factors, for instance, on the orientation of ultra-microscopic elements which crystallise in the tissues, passing from the isotrophic state of a liquid to the mesoform state or the state of a crystalline network and then to one, two and then three axes. Despite the failure of attempts to directly verify the critical period of passage from one determination to another with X-ray diffraction, it is all the more plausible, above all with respect to the auditory sketch, that many instances of mirror-image splitting, of the kind found in crystals, can be produced – as if, Harrison says, 'the transplanted tissue already knew what to do'.[23] Bernal has emphasised more generally that liquid crystals or mesoforms can easily play the role of proto-organs. Since they are liquid, on the one hand, they are not impenetrable and allow for chemical reactions and continual substitutions of molecules; on the other hand, having a structure that they conserve, they easily pass into the complete structuration of true organs.

ORGANISMS AND CRYSTALS

And yet, we should not entertain too many illusions regarding liquid crystals any more than we should concerning the dynamic forms of the 'soap bubble' or 'redistribution of electrical charges within a capacitor' kind. The completely developed organism in no way resembles a crystal. It is possible, and even likely, that the primitive stages of organic formation are indistinguishable from crystalline formation, at least if we take into account their thematic progress towards a typical structure. But the destination – a leg, an eye, a kidney, a nervous system, not only ordered like a crystal but structured like a tool capable of working – is evidence that the theme is not only an order. There are crystallisations in organisms. More than this, the spicules[24] of diatoms, Radiozoa and sponges are often genuine crystals with respect to their texture, as inspection under polarised light demonstrates. But it is extremely striking, as D'Arcy Thomson remarks – still on the lookout for what, in the organism, can be explained by structurations of the physical kind – that the exterior form of these spicules does not conform to the orthodox contour of crystals.

The forms of spicules in *Foraminifera*, composed of calcium carbonate, 'resemble none of the forms of calcite crystals; they seem to have been sculpted *in* a crystal; they are in fact constrained crystals, crystals growing in an artificial mould, so to speak'.[25] Organisms belong to very different groups, using quite different chemical materials (silicon, carbonate, strontium sulphate), but they form spicules or skeletons whose forms are almost identical, proof that their total forms do not result, as in crystals, from the forms of molecules. The dodecahedral or icosahedral shells of certain Radiozoa constitute 'impossible' crystalline forms. Are the 'artificial' mould and the antagonistic force of the forces of crystallisation not simply the superficial tension or surface energy between the crystal and the protoplasm of vacuoles and cellular intervals? In fact, as K. C. Cole, E. N. Harvey and others have shown, the forces of surface tension are, in quantitative terms, completely insufficient; they play a very minimal role relative to an existing, well-established skeleton; they cannot explain this skeleton any more than the film of soapy water is explained by the loop of wire across which it is stretched.

As Bonner has noted, 'the fact that, from generation to generation, a sponge possesses a complicated, well-defined spicule structure is entirely unexplained'.[26] But a leg, an eye or a kidney is distinct from a crystal to a much greater degree than a spicule. The fact that, at a very early moment, a meso- or para-crystalline network serves as their canvas explains them just as little as the thread of a tapestry explains its picture. An homogenous order, the monotonous sorting of identical constituents, cannot explain an articulated order whose every part, generally dissimilar from each other part, is mutually 'signifying' and whose formation, furthermore, is articulated and thematised in time as much as in space.

Above all, the contrast between the organism and the crystal has the value of clearly showing the nature of the dissimilarity between organic and crystalline formation. The growth of a crystal essentially operates according to step-by-step forces between molecules, arranging them closely together as quickly as they are attracted by the largest number of facets. There is no need to appeal to a total, dominating force in order to explain the resulting form of the crystal.

The same holds for systems produced through surface tension or through the equilibrium of a field of forces. The form of a catenary,[27] a soap bubble or a magnetic field constitutes a 'one', but this is only the unity of a result, calculable in general according to the principle of least action. The fundamental equivocation of innumerable theories of totality and Gestaltism involves the confusion of the two kinds of possible unity of a system: unity through step-by-step horizontal interrelations between elements, and unity through vertical thematism, through hierarchised liaisons. The model of this unity can be neither the crystal, nor the soap bubble nor the completely constructed and functioning machine; it can only be an active assemblage according to a theme, which makes elements disposed to each other converge increasingly well, but always according to a simultaneity which would be ideal if not actual and would then allow for a functional cycle without being reducible to it.

At the start of development, the role of step-by-step 'orders', of gradients and polarities, proves the presence of a 'transversal' unity. Gradients are for hikers and engineers. No clump of soil knows anything of gradients; it only knows of the gravity that keeps it in place.

CYCLES OF REGULATORY FUNCTIONING

Contemporary biologists, above all in the wake of the rise of cybernetics, are tempted less to reduce the mystery of morphogenesis through recourse to the models of the 'soap bubble' or 'crystal' and are instead tempted by the 'regulatory functioning' model. All functioning is cyclical in the sense that aside from wear, it can return to its starting point – thus a wheel, a pantograph or a trap (for example, a spring-loaded mousetrap). The trap, after having been used, can be reset and used again. In this last case, the cycle is not autonomous, and it requires human intervention in order to reset the trap so that it can function. But the cycle can be made autonomously dynamic – for example, a gas-operated machine gun or a combustion engine which controls the opening and closing of valves, the circulation of petrol and the ignition of the spark. If we go one step further, the dynamically autonomous cycle can be made auto-regulating through a so-called feedback* system, through an auxiliary circuit which registers the effects obtained in the functioning of the principal cycle, and reacts, according to its assembly, by reintroducing these effects as a regulatory cause in the principal cycle. Thus we have the thermostat, Watt's steam engine and so on. The auxiliary cycle can even register the effects produced by an external object, can actually receive 'information', and is capable not only of regulating functioning but of pursuing a certain effect-result: thus the cannon equipped with a corrective firing radar on a mobile base.

A cycle of this kind appears to reconcile the modality of relation and causality that belongs to the step-by-step with the modality of action and unitary liaison. In the tracing of the circumference of a circle, every cause is at the same time an effect, as its outline is traced: each point is at once origin and end, and every local disruption modifies the whole. This was already the case in dynamic systems as banal as the soap bubble. But here the new fact is that total action aims at a defined result, one to which it appears subordinated. Such a cycle easily gives the impression of a sort of conscious, finalised and intelligent 'survey'. A unitary theme appears to dominate functioning and to subordinate both energetic expenditure and articulated deformations of structure.

The impression is even stronger with respect to so-called 'ultrastable' systems, such as Ashby's homeostat,[28] in which many circuits in feedback* are mutually coordinated and appear capable of seeking out and even improvising not only an equilibrium, despite exterior disruptions, but various means for attaining this equilibrium.

Cycles of regulatory functioning are certainly very important in physiology, psychology and sociology. The organic life of an adult consists of many homeostatic systems: the regulation of sugar, water, calcium and temperature all operate according to slow feedback* regimes, putting into play the autonomous nervous system and the endocrine glands. Psychological behaviour is, in large part, also regulated in the same fashion: our actions are guided by the rapid feedback* of the central nervous system and by informing sensations. In any case, behavioural feedback* is subordinated to organic homeostases, need serving as an 'amboceptor'²⁹ between two sorts of feedback*: we experience pleasure when eating, and we seek out food while we are hungry.³⁰

MORPHOGENESIS AND CYCLICAL DEPLOYMENT

But it is clear that a cycle of functioning cannot explain morphogenesis, that is, the *deployment* of elements of a cycle. Once the deployment is effectuated, each element of the cycle can only act by mechanically pushing on neighbouring elements. But it is impossible to imagine that this mechanical causality provides us with a means to explain morphogenetic deployment.

Once an automatic chain of fabrication has been deployed in a factory, it is true that the machines themselves all but fabricate other machines, but again they too will have had to have been deployed already. Now, embryological formation is a *deployment* of organs of homeostases and organic automatisms. The need for a 'verticalism' here is clear and all the more so given that morphogenesis deploys elements of a future cycle in distinct and remote areas.

The primordia of the pancreas, secretor of insulin; of the liver, stockist of glycogen; of the muscles, consumers of sugar; and of the nervous system and organs of taste that facilitate the search for food are all developed simultaneously and in relative independence.

The unaccommodating eye³¹ of the snail, for example,³² possesses a spherical and rigid lid. In order for an accommodating eye to be produced, this lid must soften, enclosing itself, through filaments, into a softened capsule. These filaments must be attached to the ocular wall at the point where an articulated muscle can take its bearings from a point in front of it and which, in contracting, can shorten the filaments and dilate the capsule. This muscle must be controlled by a nerve which itself must be controlled by a centre (quadruplet bodies) to which signals from the optical nerve's fibres are sent and without which accommodation regulated through feedback* according to the distance of vision would not be possible. The deployment of this chain of 'amboceptors' could only come about vertically. We cannot say that it must come about instantaneously and miraculously. In fact, it is perfected through a long evolution. But we claim that this perfecting must bear on the *whole* of the cycle. *The fact that it functions, once formed, through step-by-step actions implies precisely that it cannot be formed in this way.* To maintain the contrary is to assert – to recall a remark that amused André Gide – that the inventor of the button would have had to have met the inventor of the buttonhole.

An adult individual and even a young child – since they already possess all of the organs – is capable of adapting just as a bird is capable of flying given that it has wings and a nervous system capable of an adequate level of feedback*. Those who try, even today, to assert that the bird 'flies *because* it has wings' must not forget to add that wings only exist after the being itself makes them according to a 'vertical' deployment, one which takes up in its turn a whole specific morphogenesis.

SPECIFIC MORPHOGENESIS AND INDIVIDUAL MORPHOGENESIS

If morphogenesis is considered not only in terms of individual formation but in the formation of the species, if it is followed step by step, from one species to another, according to their likely filiation and the perfecting of organs, its thematic aspect becomes apparent. The cycles of functioning can be seen to progressively expand, annexing auxiliary cycles, progressively extending their control and, in a word, being perfected in the very manner of a human technical invention – always according to their direction [*sens*] and their global output. 'Idle' organs, not part of a cycle, are rarities, explicable as the residue of previous cycles. So-called orthogenetic theories in the psycho-Lamarckian sense, and philosophical conceptions of the 'creative evolution' type, account for this vital aspect of the course of evolution – what, logically, leads to an interpretation of individual morphogenesis in each generation – as the common work of memory and invention. Not that each individual embryo must repeat these ancestral forms – but it is at least logical to think that individual morphogenesis must be caught up in the morphogenesis of the species in order for the conduct of the formation of organs to be influenced by the manner in which organs were invented in preceding individual morphogeneses.

This is, moreover, what we actually observe. The first stages of development are strikingly similar across the most varied levels. Primitive themes and organ schemata appear entirely exposed in the embryogenesis of a higher animal. From the unicellular state, it then passes to the state of a spherical cellular colony before an axis of bilateral symmetry, a head and a tail, a neural tube, a digestive tube and segments appear, and so on.

In evolutionary biology since Darwin, however, there is a general refusal to consider the morphogenesis of species as a thematic invention. For contemporary Darwinists, this morphogenesis of species is explained as an accumulation of fortuitous mutations. An approximate harmony of organisms results, in a completely negative fashion, through the elimination of mutations incompatible with survival. It is not our aim here to discuss Darwinism and selectionism.[33] We only wish to note what individual morphogenesis comes to from this perspective. It clearly poses a very embarrassing problem for selectionism. If each new being, in each new generation, were produced through a sort of tracing, everything would make sense. It is *possible* that it happens a little like this in the virus – since it reproduces by division, all structural mutation is naturally transmitted. But it is not at all the same for multicellular organisms, which must reconstitute adult forms in the course of a long embryogenesis and which do so, as we will note, on the basis of very schematic forms. If, for example, a human being is a simian *plus* mutations, why is the embryonic human less than a simian and even less than a fish or a vertebrate? If mutation is, by origin and nature, completely mechanical, entirely 'structured' in the properly spatial sense of the word, how can it remain virtual during the first stages of embryogenesis?

GENETIC MUTATIONS AND ATOMIC TRANSMUTATIONS

The genetic and chromosomal theory of mutations is meant to respond to this difficulty: the chromosomes of the cellular core, formed through the piling up of genes, must be conceived of as key to both specific evolution and individual morphogenesis. The adult form is strictly the function of the genes of the fertilised cell from which it arises; small individual variations are due to reactions to the milieu and are not transmissible. Since the gene is quite close in size to a virus, with which it shares a number of traits, we can say that in its conception, however complex an adult organism might be, it remains a sort of envelope-effect of the chromosomic kernel of its initial cell; its forms are only a vast, directed amplification. True reproduction, true morphogenesis takes place at the moment of the duplication of genes in the organism of its parents, and ontogenesis is only a formality. We are brought back to the case of reproduction by tracing.

According to this theory, the visible forms of an adult multicellular organism, for instance, a vertebrate or a human being, are comparable to the envelope of electrons around an atom, where the genes play the role of the atom's nucleus. If an atomic nucleus is able to be modified by a particular bombardment, or, more precisely, if the charge of the nucleus is modified, the modified nucleus is immediately enveloped, by capture or emission, by a lining of electrons corresponding to its new charge. The comparison is all the more applicable in that for many biologists and an even greater number of physicists, the biological mutation of genes is caused, like the transmutation of a simple chemical body, by exposure to radiation. According to this hypothesis, the biological phenomena – initially invisible (except when there were serious burns or death) but present at the germinal level – of the victims who survived the atomic explosion at Hiroshima were of the same kind as the physico-chemical phenomena that constituted the explosion and thus prolonged it. The nuclei of the uranium split in two through nuclear fission and after this fission changed their chemical kind, being reconstituted as atoms of barium. The Japanese people who were subjected to this exposure to radiation, apparently bearing no serious wounds, may have been submitted to genetic mutation in their germinal cells. These mutated genes, in reproducing, guide the morphogenesis of the subsequent generation in turn, producing organisms of a new kind, giving rise to new biological varieties. The transformation of Japanese population 1 into Japanese population 2 takes place more slowly and less noisily than that of uranium into barium or lead, but it is of the same kind.[34] The organisation of the group of atoms constituting the genes entirely governs the form of the organism, and the mutations of this group of atoms, and the various leaps from one state to another possible state of this group, entirely governs changes in the form of the species. Since the beginning of life, solar or cosmic radiation has played the same role played by the atomic bomb for the inhabitants of Hiroshima.

THE GENETIC THEORY COMBINED WITH THE THEORY OF FUNCTIONAL CYCLES

This surprising conception is rendered a little more plausible when it is combined with the theory of self-regulating cycles. There, genes do not direct the construction of each innumerable detail of the adult structure. Rather, and in the way that machines built to construct other machines function, genes only control the deployment of the primary auto-regulating cycles which are then charged with all secondary adjustments. This combination of the two theories is clearly indispensable in order to make sense of the fact that the stability of organisms is not explained by the inert perseverance of a structure and that changes to the organism in the course of evolution do not resemble the modifications of a block of marble receiving the chisel's blows at the hands of an invisible sculptor. Genes are stereotypical molecular edifices – up until the next mutation – but they fabricate systems which are themselves 'open' and supple, possessing a dynamic rather than a static equilibrium.[35] Genes, in sum, guide the fabrication of a kind of automaton akin to Ashby's homeostat or to the computers used to 'automate' a factory. Furthermore, they never fail to provide 'assembly instructions', in the form of a duplication of their own structure, to the machine that they construct. These allow the machine to reproduce the same operation over and over again until the exposure to atomic radiation, which modifies the 'instructions', modifies the entirety of the assembly line.

VON NEUMANN'S ARGUMENT

Von Neumann[36] has shown that there is no contradiction here: 'The problem of self-reproduction can then be stated like this: Can one build an aggregate out of such elements in such a manner that if it is put into a reservoir, in which there float all these elements in large numbers, it will then begin to construct other aggregates, each of which will at the end turn out to be another automaton exactly like the original one?'[37] At any given moment, a living species is thus equivalent to an assembly line in a factory in which automata are fabricated, a factory where 'automation' would be so well perfected that not even a single worker would still be needed. And the ensemble of species, in geological time, is equivalent to an assembly line whose 'instructions' would be accidentally modified by it, and which would begin from a primitive state or with a cycle of functioning as simple as a chemical reaction.

There is, then, no logical contradiction. We see why it would be no less possible to introduce 'automation' into a factory manufacturing robots than into an automobile factory, and why a machine could reproduce instructions for assembly (in the form, for instance, of a perforated tape) that it would automatically introduce into a machine capable of manufacturing according to these instructions, a machine that could in turn have been manufactured according to the instructions it receives.

THE ROLE OF GENES IN MORPHOGENESIS

But while it is unassailable from a logical point of view, von Neumann's reasoning rests on a false biological hypothesis, namely, that genes direct individual morphogenesis in the same way in which 'instructions' and 'programs' completely direct the functioning of a calculating machine. Not even the most optimistic of the experts in genetic theory would dare to affirm today that embryonic formation is *directly* subject to genetic control. According to recent research,[38] genes act on morphogenesis through the intermediaries of often replaceable chemical substances. At times, they seem to directly produce organic substances, for instance, the pigments or enzymes whose absence can trouble structuration and whose presence allows or evokes, as with a hormone or an inductor, certain developments. It is in this way that genetic constitution determines sex, at least so long as no other hormones or inductors are experimentally introduced to modify this determination. But to *evoke* is not to *inform*. As long as genes act through the intermediaries of relatively simple and replaceable chemical substances, there is no question that they could be equated with the 'instructions on a perforated tape' communicated to an automaton. The continuation of genetics in morphogenesis no longer subsists in the minds of biologists other than in the form of a kind of residual, in fact futile, hope. In taking this diminished hope seriously and presenting it so crudely, mathematicians like von Neumann and physicists like Schrödinger have done nothing but underline its inconsistency. It is clear, on experimental grounds, that the absence or mutation of a gene can *trouble* development, like the absence of a material component or the modification of a tool can trouble the construction of a house. But it is impossible to conclude from this that the *presence* of this component or tool explains construction. We cannot maintain that the progress of morphogenesis, from the virus-molecule to the human being, is explained by an accumulation of errors in the duplication of the 'instructions' in the automated manufacturing of an automatic machine by an automatic machine.

INDIVIDUAL MORPHOGENESIS AND SPECIFIC EVOLUTION

The effort to relate individual morphogenesis to the morphogenesis of the species on the basis of genetics has resulted in an undeniable failure. Nevertheless, the inevitable solidarity of evolutionary theory and theories of embryogenesis, and the fact that embryological theory must take primacy over evolutionary theory, has always been clearly understood. If the formation of a new individual cannot be explained on the basis of genetic 'orders', how can we continue to allow the claim that the formation of species is explained by the accumulation of solitary genetic mutations? The evolution of a species is only, after all, the ensemble of individual formations in both species 1 and then species 2. What is false for a given generation cannot be miraculously true for the totality of all subsequent generations. The necessary solidarity of the two morphogeneses must therefore be interpreted in a completely different manner; we are led to another type of solidarity by invention and memory, the same active thematism which explains reproduction by its mnemic aspect, and the progressive creation of a new species through its inventive aspect.

Drawing on the work of Garstang and Bolk, de Beer has forcefully emphasised the mythical character of the morphogenesis of a species vaguely imagined as distinct from and as 'cause' of the unfolding of individual morphogeneses, which would in turn be restricted to its recapitulation.[39] On the contrary, it is much rather the case that individual morphogenesis must invent specific modifications: 'Ontogeny does not recapitulate Phylogeny: it creates it'.[40] If neither the analogy with tracing *nor* the invocation of errors in tracing can be taken as acceptable interpretations of ontogenesis, there only remains the solution guided by an analogy borrowed from aesthetic creation, that of 'theme and variations'.

A simple comparison, no doubt, and science is not done by comparison. But it is equally anti-scientific to refuse any value to this comparison since to do so would lead us to once again concede the convergence, or the miraculous continuation, of two modes of development whose principles yet remain with nothing in common.

MECHANICAL EVOLUTION AND CONSCIOUS EVOLUTION

If the development of the human being as an organism and the development of human culture obey completely different principles – the one being the blind result of copying errors and genetic accidents, the other being thematised and consciously oriented – how can the one arise on the basis of the other and prolong it? 'With man', wrote J. Huxley, 'a new method of evolution has appeared, as different from the purely biological method of the natural selection of self-reproducing variants as this latter itself differs from inorganic cosmology.... Evolution is at the point of becoming "internalised", conscious, and self-directing'.[41] We freely admit that it is dangerous to want to discover a unity in the processes in nature too quickly. We must expect that we will not be able to completely assimilate cultural and biological development. But this sensible rule must not make us forget an even more decisive rule: to not derive one incompatible nature from another, and to only believe in miraculous emergence as a last resort. Now, this 'internalisation' of development and evolution, this conscious taking charge of evolution spoken of by Huxley, would indeed be a miraculous animation, the magical metamorphosis of a marble statue into a living woman. If the human being is an automaton produced through a chain of automation, like a statue without a sculptor, it can even less animate itself or, being a manufactured product, become an artist. It is difficult to think that blind nature has constructed, without consciousness, a consciousness capable of dominating it.[42] It is difficult to agree that the organic tools human beings consciously make use of in the course of behaviour dominated by elaborated themes were constructed outside any consciousness. It is difficult to agree that the nervous system, the instrument of evolutionary self-organisation, was first the product of billions of mutations resulting from mechanical copying errors. The cycles of organic homeostasis, behaviours and techniques cannot be absolutely heterogeneous; the deployment of their elements cannot be subject to unrelated principles. The 'vertical' development of cultures and techniques – that is, in spite of innumerable coincidences, encounters, 'horizontal' disturbances, their effort to improve themselves according to a collective theme – is the natural consequence of the vertical and thematic development of organisms. A functioning cycle or an organic homeostasis passes from a simple to a more complex state – or, at times, as in neoteny, returns to a simpler state in order to become more complicated in another direction – but it is always the cycle as a whole that must be given. Primitive birds did not fly in the same way as the birds of today for which, as J. B. S. Haldane has shown, proprioceptive feedback* stabilisers have had to develop in order to replace the long rigid tails of the pterosauria and the archaeopteryx in balancing flight. But changes in organic morphology belong to the same order as changes in technique: Blériot's plane was different from a Super Constellation, but it was a plane, and it flew.[43]

Chapter 2

From the Molecule to the Organism

If we concede the thematic character of individual organic development and, furthermore, the identity of individual and species morphogenesis, we would seem to be led into pure mythology. Where does this formative theme come from, and at which moment does it intervene in processes of physical functioning? The origin of 'vertical' formation in individual development is an egg, or a specific 'living' cell. While the respective forms of the egg and the adult differ dramatically, and while modern biology is a long way from the earlier theory of preformationism, biologists can be reassured by insisting, as Jean Rostand does,[1] on the fact that the egg possesses an extremely complex organisation. Its architecture does not prefigure the future organism, but it is already an organic architecture such that development passes from the organic to the organic. The augmentation of complexity is mysterious, but at least it is not the passage from one mode of being to another. But where does the first cell come from? The morphogenesis of a species as a whole certainly does not originate with a completely constituted cell but rather something which must have resembled what we today call a virus, a self-reproducing macromolecule. From viruses to higher animals, thematic development seems to come into being at the level of the chemical molecule. The molecule certainly possesses a complex architecture. But what relationship holds between the complexity of an atomic edifice united by chemical valence – by bonds that seemingly function in step-by-step fashion such that the molecule appears to resemble a kind of Meccano in which pieces can be added, subtracted or replaced by others – and the complexity of a living being which has a closed individuality and 'organic' parts, that is, analogous to tools, possessing a role and a function that can be neither isolated nor substituted? It would seem that a molecule functions according to its structure. Its chemical properties are theoretically deducible from its 'developed formula'.[2] An adult organism, and above all a mature organism, also functions according to an acquired structure, but as we have seen, a young organism, and above all an organism in development, cannot be considered to be functioning by the same logic. How then can *roles-functions* come into being on the basis of molecular *functioning*? How can an individualising formative theme come into being or come to be inserted into a edifice constituted, it would seem, by bonds that are linked edge to edge?

THE VITALIST TEMPTATION

The temptation presented by vitalism or animism is considerably better justified than mechanists commonly think. It does not simply rest on the fanciful belief in a kind of vital breath which is added to the visible matter of the living being, introduced like the breath of Yahweh into clay formed into the shape of a man.[3] It rests instead on the intuition that organic forms are not of the same kind as, or extrapolations of, physico-chemical forms and that their respective modes of complexity are entirely different. Such was the scientific error, prominent in chemistry, made by the Cartesian biologists of the seventeenth century who, like the authors of Genesis, also imagined that the form of an animal could be directly shaped – if not by Yahweh, then at least by the laws of physical nature.

And yet, if a more advanced science, and the use of a microscope, instead promoted vitalism, a more advanced science yet again – with the use of an electron microscope and thus the indirect study of submicroscopic structures – has clearly lead to anti-vitalist conclusions. Nature manifests something that at first appears impossible: as if through imperceptible turns, it engenders organic forms from molecular ones. To the eyes of the scientists following these turns step by step, the invocation of vital forces does not appear necessary. To cite one authority on the matter,

> A revolutionary fact which emerged from the synthesis of organic compounds was that, in chemistry, there is no fundamental difference between living and inanimate matter [. . .] Now, morphological formation in the submicroscopic world presents an exactly similar case. Whoever had expected to find special biological formative principles, alien to the inanimate world, in these invisible regions, is doomed by the results of research into natural substances of high molecular weight to as great a disappointment as was at one time suffered by the believers in mysterious life forces which alone were deemed capable of building up organic compounds. The formative forces in protoplasm and its derivatives are no different from those operating within inanimate organic Nature. There is no evidence of the existence of novel formative principles beside the atomic valence and molecular cohesive forces and their diverse *patterns*.[4]

Frey-Wyssling adds that this should not fundamentally surprise us since, in the molecular world, the chemical properties of bonds and formative properties are indiscernible. In a molecule, all morphological change implies a chemical change: matter and form are strict correlates.

The organic world simply manifests the same great fact: form is inseparable from matter. Living matter only ever appears as formed, just as the benzene molecule only ever appears, as matter, in its well-known hexagonal shape. Benzene is not an amorphous matter that comes to be 'informed' by the shape of the hexagon, produced like an Aristotelian form. It is this form itself, which is in turn derived from the modes of bonding of carbon and hydrogen. In the same way, biological forms arise without hiatus from molecular morphology. We do not yet know how the specific visible forms of cells are derived from molecular networks, but doubtless relations exist between molecular morphology and the morphology or organic morphogenesis already found in the chemistry of enzymes or the asymmetrical synthesis of organic compounds. A protoplasm is not a liquid in which isolated particles float randomly, pell-mell; it has a flexible, subsistent structure in which active centres can only exercise their functions if they can rely upon the network in general, 'like leaves attached to the branch, and not leaves detached and tumbling in the air'.[5]

However inconceivable it may appear, the facts seem to show this transition between a structure of the 'molecular network' type and a structure of the 'tree' type in which each of its constituents, even though they are assembled by adjunction and repetition like leaves on a branch, plays a role in the whole and cannot be detached from it.

FREY-WYSSLING'S PRINCIPLE

We find here a sort of intellectual scandal. If the zone of 'turning', the passage from macro-molecules to elementary organic forms, is left unexamined in favour of considering the destination of organic morphogenesis (the adult animal), the difference, even at the strictly morphological level, in fact lies between forms of the 'molecular network' type and forms of the 'organ' type such as the heart, the eye or the lung. Frey-Wyssling recognises this when – in emphasising the fact that complicated chemical processes like the morphogenesis of living beings are not controlled by a vital principle but only consist of innumerable reactions or bonds, each of which is accessible to a causal investigation – he adds, 'Yet no simple mechanistic interpretation can account for their delicately attuned harmony and their purposiveness [. . .] The active centres of the protoplasmic network are arranged according to a flexible pattern*, which seems to be guided by a purposeful, coordinative impulse'.[6]

The only possible solution is not to renounce the principle of verticalism but, on the contrary, to extend it to chemical morphogenesis itself. This is what Frey-Wyssling does, however implicitly, when he states what he calls an 'axiom' but which he would have been better to call a principle: 'The supreme axiom of cytology, namely, that all cells derive from their like, applies equally, though in a wider sense, to invisible submicroscopic cytogenesis: *Structura omnis e structura*'.[7] This principle, as it is easy to see, is nothing other than the principle of verticalism in another form.

MICROPHYSICS AND BIOLOGY

But this interpretation of micromorphology requires commentary and elaboration. We insist, first, on what could rightfully be called a misfortune provoked by the overly quick intelligence and excess of humour of the kind we find in N. Bohr, P. Jordan and Eddington, or, if you prefer, by the slow wits and poor humour of their commentators. The first, N. Bohr, having clearly understood that the structuration of the atom cannot be explained by the laws of mechanics and classical physics, themselves secondary to and derived from the principles of microphysics, has grasped the interest of this great novelty for the interpretation of life. Biologists, despite the great desire they have for this discovery, do not manage to connect organic forms to ordinary 'formations' in physics. But this failure of classical physics is entirely endemic. Classical physics is only concerned with crowd phenomena. Microphysics, on the contrary, leads naturally into biology. If one begins with the individual phenomena of the atom, one can, in effect, move in two directions. Their statistical accumulation leads to the laws of ordinary physics. But while individual phenomena are complicated by 'systematic' interactions, they maintain their individuality. Even though from the heart of the molecule to the macromolecule and the virus and then to the unicellular organism everything is subordinated to crowd phenomena, however large they become, they remain in this sense 'microscopic'.

MICROPHYSICS AND 'FREEDOM'

Unfortunately, instead of remaining at the level of this correct and general account, N. Bohr, P. Jordan and above all Eddington – leaving the careful analysis of this passage to the on-going work of micromorphologists – become caught up in the attention given to the problem not of morphogenesis but of 'determinism and freedom', entertaining a direct comparison between the individual atom and the individual organism and even, to be more precise, the individual atom and 'the person who hesitates at the moment of decision' ('The bachelor who asks himself whether or not to remain single', as Eddington quips). What interests them, and even more their interpreters, is the 'gap in determinism', the lingering possibility that certain 'key atoms'[8] would manifest the action of a directive idea, an individual will. Such a direct confrontation calls for critique even though it already does not seem particularly serious. R. Poirier is surely right to say,

> It is hard to see how human freedom intervenes on the infra-atomic level in the way that someone might move the hands of a clock with their finger. . . . Once it is construed in terms of an homunculus, the question of how it could act on matter, or, incidentally, how, such a decision made, this action could find a place in classical physics, makes little difference. But all of the difficulties, all of the antinomies of freedom, concentrate on the existence and nature of this little being who is only an image in miniature of all psycho-organic organisation and reprises all of its paradoxes.[9]

We can add that even the 'solution' to the problem of 'freedom', or of the a-causality of the living, only renders the morphological problem more acute. The translation of microphysical a-causality into the register of the visible organism implies, without explaining, the existence of complicated structures in order to maintain and amplify the individual microscopic phenomena and to translate the putatively free 'quantum leap' onto the level of decisions on a human scale.

MOLECULES AND MICROBIOLOGY

These are forceful critiques, but only with respect to a lamentably inadequate way of expressing a correct idea. Leaving aside the confused determinism-freedom debate, let's recall the continuing patient work of micromorphologists on the structure of gels; on the delicate morphology of the cytoplasm; on macromolecular fibres and contractile proteins (keratin, muscular fibres); on genes, viruses and the modes of their self-reproduction; and on the bonds and changes in chemical bonding which express their changes in form.

> Furthermore, above all in England and the USSR, better-armed biologists are today attacking the problem of the origin of life and trying to locate the stages that must have led from certain molecules – or, as Bernal will say, from sub-vital structures – to viruses and bacteria. It is striking that the specific evolution of bacteria and viruses is, like their form, a curious intermediary between chemical and organic change. J. B. S. Haldane has emphasised[10] the fact that bacteria and viruses can make use of and copy parts of other viruses, even those of different species. A bacterium can incorporate a large nucleic acid molecule from another strain, or indeed from another species. These molecules are thus reproduced in their new environment, and the synthetic bacterium is capable of producing a new type of enzyme. Thanks to the work of Taylor, we know that a pneumococcus can incorporate substances from two different sources and constitute a hybrid molecule through a process analogous to the crossing over* of chromosomes.

ATP (adenosine triphosphate) molecules, considered by most chemists to form at a very early stage, behave in a semi-vital fashion, being able to replace a lost part according to a kind of metabolism. J. B. S. Haldane amuses himself by treating the situation in which two biophosphoric molecules give rise to a triphosphoric molecule – along with a rapidly decomposing residue of adenylic acid, here the analogue of a polar globule – as a sexual process.[11]

THE HISTORICAL SUBSISTENCE OF ORGANISMS

Even when posed in less hypothetical terms, such attempts to deploy these faulty arguments in no way resolve the problem of the passage from one type of form to another, from chemical structure to organisation, or from one type of subsistence to another. It is not always possible to see how chemical form – which appears to subsist through structural inertia in the manner of a construction using Meccano – can, however invisible the transitions might be, change into another form whose mode of subsistence will be completely different, dynamic but also historically inflected, that is, not only dependent on bonds that are themselves timeless but on past moments in evolution. While the structure of a molecule is independent of history for a chemist, depending only on timeless laws of structuration, the form of an organism or an organ, in spite of the transitional case of 'synthetic' bacteria, is maintained and enhanced throughout an evolutionary lineage. Without doubt, a human hand owes its *present* material consistency to the chemical solidity of bone, or to the bonds of colloidal protoplasm – but the same is true for a wing or a fin. The question is knowing how the same chemical cohesions are employed in such different biological forms, forms which subsist over time while being progressively modified. It thus appears necessary that, at the level of sub-vital elements, something is added to chemical bonds which will be the primer for both this historical subsistence and properly organic forms: as J. D. Bernal remarks, 'Life, even at its most primitive, is more than a system of sequential reactions. Characteristic material structures, including nuclei, cells, are formed and involved in indissoluble relations with both the chemical phenomena that produce them, and with their evolutionary origin'.[12]

THE DYNAMIC SUBSISTENCE OF MOLECULES

But there is another chapter of contemporary science that must not be forgotten since it provides both a solution and a warrant for micromorphology and support for the hypotheses of Bernal, Haldane, Pirie, Oparin and Dauvilliers on the chemical origin of life. This is the interpretation, advanced by the new physics, of bonding and chemical structure. As long as the examination of bonding and chemical structure takes place according to static, geometric and mechanistic schemas, as it has for the past half a century, as long as the comparison is between chemical structure and a 'construction' whose pieces, in themselves distinct, are extrinsically reunited, the possibility of a passage to forms and organic formations will clearly remain unknown. But if we adopt the current understanding of organic chemistry, the rapprochement at work with organic morphology suddenly leaps out. Quantum chemistry and the theory of chemical bonds drawn from wave mechanics conceive of chemical substances as characterised less by a structure than by an ensemble of structural states or *structural behaviours*.[13] Chemical morphology is first morphogenesis.

> Carbon, for example, is no longer conceived – if it ever was by the chemist – as a small solid with the shape of a tetrahedral pyramid that could be stacked alongside other small pyramids. The 'tetrahedral' hypothesis only implies 'the quadruple orientation of possible syntheses'. The form is virtual and depends on forces of composition. Valencies and chemical bonds do not signify the 'existence of hooks' or any other latching mechanism. The structural concept of valency has ceded its place, in the first instance, to dynamic analyses of bonding – the electrons involved interact according to their 'state' – and then to probabilistic analyses of interaction. Carbon is not quadrivalent but rather bears 'a very strong probability of tetrahedrisation' when the four carbon-proximate atoms are identical. The carbon atom, and this is true for every other body, is not a structure but represents a *structuring activity*, an activity 'which consists in filling space with increasingly numerous and delicate supplementary conditions'.[14] The properties of valence possess the characteristic of virtuality. The molecule as such – in opposition to the atom – already has a structure in the sense that the nuclei of the atoms which constitute it form a relatively stable pattern*, one that can even in certain cases be photographed using the technique of liquid crystalline networks. But electrons engaged in bonding and interaction are not localisable. Today, the formulas developed in chemistry, conceived *more geometrico*, have been replaced with electron density maps which represent, depending on the profile of the level, the means of structuring comportment. A defined state is only an abstract instant in a process of continuous formation. Two regions, neighbourhoods in an electron density map, conjointly 'structure' according to their energy of interaction and resonance. In benzene, for instance, the electrons involved in the double bonds are clearly even less localisable than electrons in simple bonds. Double bonds – which are represented as alternating with simple bonds in Kekulé's diagram[15] – must be symmetrically divided in every molecule in order for them to be stable. Static representations are only sketches of skeletons. In order to grasp reality in full, this reality must be understood in terms of the energetics of electrons. The molecule is a domain in which energies are exchanged, in which energy structures itself, in which a structural state is 'chosen' from among an essential multiplicity of possible states.

With Bachelard, we can speak of the anatomical and physiological aspects of the molecule. In order to avoid any equivocity, a 'morphological aspect' could be distinguished from a 'morphogenetic dynamic' of the molecule since 'physiology' is here formation and not functioning and since it is the activity of liaison which establishes the pattern* that it actively constitutes. From physiogenesis we pass, in the molecule as in the organism, to physiology properly speaking; from structuring comportment we pass to comportment which depends in great part on constituted structures. But this is only a limit, never attained, for morphogenetic forces are always at work. In other words, 'verticalism', the dynamic deployment of functional elements, is primordial, *in chemistry just as in biology.*

FORMATIVE BEHAVIOUR IN CHEMISTRY

Consequently, we can now see clearly what before could only be glimpsed, the truth sought by micromorphologists like Frey-Wyssling and Staudinger, by physicists obsessed with the question of indeterminism like Jordan and Eddington, and those like Haldane and Bernal, searching for the missing links* between chemistry and the evolution of life. The same truth was also at stake in the often confused groping of the contemporary adepts of vitalism, or rather pan-psychism, who keenly sensed the logical impossibility of life and organic forms emerging from molecular forms, such as they were conceived of at the end of the nineteenth century, and who, given their poor grasp of contemporary chemistry, thus felt obliged, though they were unable to deploy a vital principle or unearth an entelechy, to attribute what E. Boutroux calls a 'source of all life'[16] to matter in general, or otherwise to merely engage in empty rhetoric. Chemistry and contemporary biology – which share nothing with the vague, renewed vitalism of the Romantics, or the brute application of quantum theory or indeterminist physics to poorly understood genetics in the style of today's physicists – allow for quite precise possibilities of association.

We have been able to account for both organic morphogenesis and molecular morphogenesis without becoming involved in the freedom-determinism opposition. The phrase 'the freedom of the embryo' means very little, and 'the freedom of the molecule' means even less. The true opposition in both domains is instead found between 'functioning' and 'formative behaviour, the key point is that 'formative behaviour, is the only suitable expression, in chemistry as in biology, and that 'functioning' is, in both domains, always secondary and derived.

The formative behaviour of an atom or molecule is not, strictly speaking, 'thematic', if the word is taken in the more familiar sense of 'signifying', as it is in organic morphogenesis. But what matters above all is that it no longer be conceived of as 'positional', as it is when describing the movement of a mass or a machine. The formation of a molecule or an atom is unlike a mountainous folding or a sedimentary deposit. The atom has a typical form, virtual or actual. A 'bombarded' atom does not resemble a bombarded house or a car accident. A nuclear bombardment produces a *typical*, not *random*, result – a new chemical being, whether by fusion or fission. The difficulties confronted by determinism give rise to a completely positive meaning: bonds or interactions are primary in the individual domains of chemistry, and it is pointless to try to explain primary bonds by analogy with a mechanical link since the latter is *derived* from the former. If structuring action cannot be explained by the laws governing an existing structure, interstructural or liaising action cannot be explained by secondary techniques based on established connections, for instance, a machine. Electron density maps, or maps of the probable intensity of bonds, represent something absolutely fundamental, allowing us to discern the mystery of chemical individuality and living individuality, of type and morphological theme.

THE VIRUS MOLECULE

We know that certain viruses are crystallisable and that others are quite probably monomolecular. The virus of apthous fever is only ten times larger than a sucrose molecule and only fifty times larger than a hydrogen atom.[17] Viruses have an anatomy and a physiology – or a morphology and a morphogenesis – terms which must no longer be considered metaphorical here. Bacteriophages, coliphage pairs, T2, T4, and T6,[18] all of which are even smaller than crystallisable viruses like the tobacco mosaic virus, have a membrane, a prismatic 'head', and a sort of tail or horn with which it attaches itself to bacteria and through which it empties the contents of its 'head' into the nucleic acids of its target. One does not have to go much further to imagine the recognition of a crystallisable virus and its developed formula, which, despite its size, is made of regularly alternating nucleoproteins, or of the apthous fever virus, or horned coliphage, or even larger viruses yet again, such as influenza and its vaccine, or even those whose size and appearance already possesses a cellular character. If these developed formulae are modelled on chemical formulae of the earlier kind, whose dashes and valencies evoke hooks or snap fasteners, vital behaviours – individuality, regulation of form, reproduction, heredity – become not just mysterious but inconceivable. Something, it seems, must be added to the formula, the insufflation of another world, in order to transform this Meccano into an individualised, aggressive, living being. But if we interpret the same formula according to modern bonding maps, on which the simple curved lines indicate probabilities of interaction by 'resonance' – or rather, since physicists are resistant to the literal interpretation of this term, of the partial discharge of individuality into 'neighbouring' subdomains – the term 'vicinity' too must be considered in the specific non-spatial sense of 'conjugated systems'. The physical mystery of unobservable and non-localisable bonds, known only through their effects on the spatial but also temporal comportment of the molecule, thus comes to perfectly coincide with the biological mystery of the same molecule.

63 The typical, individualised behaviour of the virus, its active and combative perseverance, appears to be of the same nature as the individualised comportment of a benzene molecule, a molecule of water or of any molecular, atomic or subatomic individualities. The temporal, and not simply geometric, aspect of bonding activities at every level prepares for the 'historical' and 'hereditary' subsistence of the virus, which is wrongly taken to be a simple molecular structure in space but which reveals their vertical 'temporalisation' such that they are capable of passing from the virtual to the actual through a kind of reproductive mnemic comportment. At first glance, it appears more economical and scientific to explain the reproduction of the virus in terms of a spatial 'moulding', but we must ask if we are not buying into a bad deal. For whatever the particular path the immensely complex process of the reproduction of the adult organism takes, the metaphor of spatial moulding is practically worthless. What is really economical is to redeploy in each case the modern category of bonding through temporalised structuring activity drawn from the level of molecular reproduction.

In bacteria infected with the phage known as 'temperate' which subsists as a prophage, this prophage, integrated into the hereditary fabric of the bacteria, plays the role of a sort of caretaker of a potential property of the cell as if it were a gene. The 'sigma' factor of the fruit fly (attracted by an awareness of carbonic oxide gas[19]), which is carried like a gene, originated as a virus – another reason to refuse the absolute separation of the reproduction of phages and the reproduction of higher organisms. What yet remains mysterious is the question of knowing why certain molecules, genes and viruses have the property of reproducing themselves while, in general, ordinary molecules do not. Why, for instance, is a molecule of alcohol added to carbonated water unable to reproduce itself, even when all of its elements are found there? Why,
64 Gamow asks, is an entire glass of sparkling water not transformed into cognac, his drink of choice, when a drop is added, as a virus completely transforms all bacteria into viruses?[20] The response will probably be able to be given when, on the one hand, the energetic conditions for the reproduction of molecules are better known and, above all, when chemistry will have been able to establish the developed formulae of virus molecules. It is highly likely, for example, that only certain types of chemical bonds involve a true systemic unity and also that only certain networks of these bonds can provide a complex molecule with an individuality capable of being imposed on neighbouring materials.[21]

THE VIRUS AND PSYCHISM

The price to be paid, if it is one, is clearly that of admitting that every molecule and even every atom is as 'alive' as a virus. An observer of this evolution inattentive to contemporary science might believe they find here a return to vague and out-dated conceptions of animism, to the imaginary attribution of a consciousness to physical matter – a miniaturised human consciousness in the form of a small demon or homunculus, the bearer of freedom, memory and intention. We think that the fear of 'verbalising' must not lead us to fear words. We must not be afraid, that is, of using the words 'organism' and even 'consciousness' apropos of molecules for it is no longer here a question of a careless use of words but, on the contrary, of an interpretation made possible by chemistry and modern microbiology concerning the reality that these words designate. Active bonding; structuring behaviour; systemic unity through themes or non-localisable types; verticalism; form and self-formation; formative instinct; absolute domain; organic domain; domain of consciousness – all of these expressions are synonymous. Wherever there is non-functional, formative activity, there is inevitably a 'for itself', self-possession, self-given form bound to itself absolutely and not constituted by secondary step-by-step bonds. Wherever a being comports itself[22] – that is, does more than function within the limits of a given structure – there is necessarily consciousness, that is, the improvisation of bonds according to a theme which is not given in space. That which functions can do nothing in itself, being just a mass or sequence whose unity is only given by others. That which forms itself or comports itself is necessarily a real, a 'for-itself'. If – with the entirety of modern physics – a purely functional conception of the atom is rejected, then, *for the same reason and by definition*, we cannot avoid attributing to the atom the status of a vision-understanding analogous to a visual sensation, or a melody-duration analogous to an auditory sensation. The negative aspect of the explanatory deficit of functioning logically finds its positive counterpart in the absolute presence of a formal unity, and a 'domainial' auto-conduction. And it is easy to verify that this definition applies equally to the molecule of water as theorised by Heitler and London, the molecule of benzene according to wave mechanical theories of mesomery, the virus, protozoa, metazoa in formation, the nervous system in the course of its comportment and finally to the field of consciousness or active perception which is 'mine' in this moment. The intuition that I have of the thematic unity of my acts, in complete ignorance of the details of their nervous and muscular actualisation, is the exact complement of what a physiologist can observe in my organism. Her observation, the complete inverse of my own consciousness, follows the detail of this neuro-muscular actualisation but it cannot grasp what provides the unity of these moments or any behavioural act. It is natural to think that in these other cases, there is an absolute presence of unity analogous to that which provides the complement to the deficit observed by an external party – who cannot account for the unitary behaviour of either a molecule or an organism, or reduce either to a pure functioning.

The mystery of life is nothing but the mystery of consciousness, which is in turn nothing but the mystery of all primary bonds, all birth of true form in morphogenesis, whether chemical or biological. A primary bond is unobservable. It can only be represented, circumscribed by probability of interaction curves. The natural difficulty of understanding the life Bergson attributed to intelligence, the fact that intelligence is only at ease with solid objects and their mechanical assembly, derives from the same fundamental law. And this law is itself drawn from the character of structural correspondence in scientific knowledge that we defined at the beginning of this work. A phenomenon becomes conventionally intelligible when the problem of a bond is taken as solved without considering the topographical arrangement of its parts. It is not the fact that something built in Meccano is solid that renders it intelligible. On the contrary, its solidity – whether of an individual piece, or of the nuts and bolts that join it to another – is itself only fundamentally intelligible on the basis of primary chemical bonds. It becomes intelligible when we concern ourselves with the arrangement of the pieces, visible in space and easy to follow in optical or tactile perception.

SECONDARY BONDS IN THE ORGANISM

Of course, not everything in the individualised domains of chemistry or at the level of the organism is a primary bond. It is clear that the articulation of the knee or the shoulder is no more mysterious than the articulations of a machine. The same holds in chemistry: *given* carbon-hydrogen and carbon-carbon bonds, the passage from methane to ethane, and from butane or aspartic acid to glutamic acid through the H-C-H relay is as clear and straightforward as the construction of a tower of dominoes. The process by which nucleoprotein chains are constructed is on the way to being just as well understood. However complex their sub-unities might be, crystallisable viruses appear to be composed of regular groupings of these piled-up sub-unities; the crystal of the virus is made in turn of these piles, according to the ordinary laws of the physics of crystals. A mass of these crystals is a long way from primary bonding and consequently from the primary mystery of the virus's individuality.

In organic forms more generally, it is easy to identify agents of bonding as secondary as glue is in the construction of boxes or cartons.

Bonner[23] has studied the case of colonial diatoms and their processes of intercellular adhesion, both through the interlocking of siliceous spines – such that the individuals, which look like long pencils lying side by side, can slide over one another, taking on the appearance of a carpenter's ruler being folded and unfolded (*Bacillaria paradoxa*)[24] – and through secreted mucilaginous substances (*Cymbella*). These bonding substances [*substances de liaison*] are secondary to the degree that cells can detach and continue their lives without any apparent damage. Nevertheless, they contribute to the production of the general form of the organism alongside primary bonds.

Certain colonies of lower organisms are little more than masses coordinated by secondary liaisons acting in a step-by-step fashion. And certain so-called social aspects of higher organisms, which are directly derived from primary instinctive comportment, are of the same sort when they act step by step – when resting in a tank, trout sometimes lay side by side in an even row. Seen from afar, they could be taken for molecules in a crystalline lattice.[25] Finally, when bonds act in an even more superficial and directionless fashion, there remain only pure masses without any genuine individuality, no longer globally structured by the statistical and secondary laws of classical physics.

We grasp, then, the bifurcation which leads from the physical world to crystallisations or inorganic masses, following vertical morphogenetic lines. We also grasp the degree to which the illusion – which sees the organism only as a singularity in the physical world, derived mysteriously and inexplicably from this world – is fundamentally false. The derivation proceeds in the inverse direction. The world is only a gigantic mass of organisms, both small and large, and what is known as the 'material' world is only opposed to the world of the 'living' because it is a mass of the smallest of organisms. The attempt to derive organisms from a secondary arrangement that supervenes over physical masses, conceived as primary, is doomed to failure. Frey-Wyssling's principle, the schema of 'verticalism', sets the record straight.

This monumental, secular error of mechanist philosophy is identical to that which, before Planck, tried to explain the physics of the individual atom on the basis of physical laws only applicable to atoms taken in great number. The converse is true: it is the physics of the individual atoms which allows us to return to the physics of multiplicities.

THE ORGANIC USE OF SECONDARY LAWS

Individual organisation always begins 'on high', that is, with a unitary theme even when it appears to be born 'down below', that is, from elements functioning according to their arrangement. The articulations of an elbow or a shoulder, like those of an assembled machine, cannot explain the assembly of this articulation. An active deployment in a domain which surveys itself, that is, possesses itself in its own form, is indispensable for what then follows – mechanisms, self-regulating circuits, sequencing and channelling of all kinds, arrangement and functioning according to this arrangement. In the living organism, furthermore, active assembly is rarely done once and for all. An individualised field continues to survey subordinated mechanisms from 'on high', compensating for failures in functioning or reorienting the manner in which a comportment is undertaken. A skeleton assembled and articulated with wire in the cabinet of a natural history museum is only a crude copy of a living skeletal system, which actively maintains these articulations and which is capable of repairing a fracture.

For this reason, it is impossible to absolutely separate the free part of an organism – which could come down to something as little as a 'key atom' – and the part submitted to the ordinary laws of physics. 'Freedom', or rather vertical thematism, is, as Whitehead said, pervasive*.

By way of example, we can consider the case of the numerous trapping organs in plants, such as the leaf-trap of the Venus flytrap. The schema of its functioning is easy to mechanically replicate, but it is impossible to compare these constructed traps with the instinctive use of traps in the animal kingdom. Both the ant-lion, which makes use of the mechanical properties of a funnel of sand but which also digs the funnel and can throw sand in the hunt for its prey, and the spider, which weaves its web and knows to rush towards its prey when it is caught, make manifest the subordinated character of mechanical functioning. The mechanical part of the trap is *inserted* into a formative, performative and reparative behaviour. On the other hand, it has been shown that carnivorous plants, whose traps are set by their morphogenesis and not, apparently, by instinct like the ant-lion or spider, are plants that suffer from a nitrogen deficiency, living in swamps where the thickness of the moss deprives the water of all usable nitrogen.[26] The construction of traps is thus, through a complex circuit that deploys and makes use of multiple physical functions, indirectly connected to a fundamental chemical need of living substances. A more complicated case of the same kind is that of the human organism, which is incapable of organically fabricating certain vitamins but acquires them through an external – industrial – circuit. In the same way, many living beings supplement a chemical deficiency through parasitism, symbiosis, the conquest of prey, etc.

Organic morphogenesis continues chemical morphogenesis and seems to put it into its service. Conversely, chemical behaviour is sometimes put to use quite directly, even at the highest level of organic life. Goldacre and Lorch have shown that the movements of amoebae, which were thought to be a phenomenon of diffusion or surface tension, are more likely to be explained by the self-folding and unfolding of protein molecules in the amoeba.[27] Like movement in the protozoa more generally, moreover, this constitutes a true comportment. A continuity can therefore be observed here, as before in the case of the virus, between chemical and biological comportment. Incidentally, contractile proteins also play many other roles, notably in the movement of vibrating cilia and in muscular contraction in general – so much so that human movement derives, in a quite direct fashion, from the contractile properties of molecules.

But we should be wary of concluding that a disincarnated intention makes use of molecular behaviour, and thereby returning to a completely nominal vitalism. Ends and means are very difficult to distinguish in an organism precisely because the constituent individualities of a higher organic individuality, with its cycles of behaviour and complicated relays, are themselves assembled and transformed into agents in one of its cycles. When, lacking a tool for the job, people form up into a chain so that a bucket of water can be sent to extinguish a fire, they are also subordinated to a total behaviour: consciousness of the system envelops the conscious attention of each person as they pass the bucket to their immediate neighbour. In all collective behaviour, constitutive individualities are at once ends and means. What is gained, in human or even animal society of a high enough level, by the adjustment of instincts or conscious calculation, is gained at the molecular level through the indetermination of the individuality of the constituents in a system of interactions. This indetermination of individuality implies an indetermination of means and ends.

This is why, for example, it is impossible to say whether the remarkable energetic properties of ATP[28] molecules or chlorophyll are utilised by the molar organism because it requires substances that use or emanate energy or if ATP molecules and substances capable of photosynthesis are primitive organisms which are subsequently associated through more complicated cycles of functioning.[29] Or rather, it is not a matter of choosing, and that both of the two hypotheses must in truth be accepted at once. A protein, or a virus with its chains of peptides and the attached amino acids, resembles a living tool – though not the pure tool *of* an organism, even if we consider the protein that has become an enzyme, or the virus integrated into a bacterium as a kind of gene. The extreme difficulty in deciding between theories of the virus as an autonomous, primitive organism and of the virus as a part of a ruptured cell is thus only a particular case of the same phenomenon.

MORPHOGENESIS AND CANALISATION

It is nevertheless necessary to distinguish, despite the practical difficulties of doing so, between what could be called the governing organic form that extends the chemical form, found in both large multicellular and unicellular organisms, and the material framework or technical machinery assembled by secondary bonds. To return to our earlier comparison, when they grow weary of passing the bucket of water along, the chain of people could instead make use of a hose, an electric pump or even fire detection systems that themselves set off automated extinguishers. In a manner of speaking, extrinsic techniques allow an organism to capture and dominate a matter and material exterior that it did not make the effort to assimilate but which it processes in bulk.

Only humans have deployed extrinsic techniques to such an extent. All the material and energetic resources of the external world end up channelled towards its ends. The planet, having undergone the greatest transformation, bears the mark of the human organism to the point that an extraterrestrial astronomer could see it from millions of kilometres away.

But the human technical revolution is only the repetition of a first technical revolution in the history of life on another scale. This first revolution provides the key to the enigma that the discovery of crystallisable viruses has posed to contemporary science: the discrepancy between the emergence of a vital property as characteristic as reproduction and the emergence of *morphological* properties habitually related to organisms. From a large molecule (non-living in conventional terms) to a crystallisable virus, there is little difference in form and one trivial in comparison to the morphological difference between a virus and a mammal or a human being. If life is determined to begin with the virus by virtue of its capacity to reproduce, it remains necessary to invoke another beginning, after the virus, that of truly organic morphogenesis. This delay in the manifestation of 'vitality' in organic morphology – reproduction itself – is what the vitalist most profoundly fails to grasp, obliging him to invoke the intervention of a vital principle into the purely chemical form of a molecular chain without organs. But how can a vital principle, if it is necessarily presupposed by the property of reproduction and is therefore supposed to absolutely distinguish the virus from the chemical molecule, not immediately induce a great difference in form between the molecule and the virus? How can an 'entelechy', supposedly the bearer of a form to be imposed on matter, modify the form of the organic molecule so insignificantly? How can it not first come to bear on the capacity for reproduction? And yet, it would be still more difficult again to situate the virus and its mode of production outside life since the particularly complicated phenomena of reproduction in mammals or human beings also incorporates, at a fundamental level, a duplication of viral type: that of genes.

THE IMPORTANCE OF PLUMBING

The enigma remains insoluble if organic formation, starting with the simplest crystallisable virus, is conceived in terms of an elementary 'framing' technique or 'canalisation', allowing for the bulk processing of exterior matter. In a remark that goes beyond the most ambitious dialectics, Lichtenberg writes, 'The most important things come about only because of pipes. The proof? Our reproductive organs, the writing quill, and the gun. What is man, if not a confused tangle of pipes?'[30] Canalisation, the tube – a rigid structure along which a crowd or amorphous fluid is carried, ready to be put to work, which allows for the imposition of order on a 'molar' dynamism, which allows for the regulation of circulation, and, should the pipes loop back on themselves, for recurrent cycles – is in effect the schema of almost all technique. It would be simple to show that human civilisation is founded on plumbing and irrigation, on the evacuation and adduction of energy, and that barbarism consists in letting the plumbing deteriorate. It would be simpler again to show that the organism only functions thanks to kilometres of tubes – for the circulation of blood, respiration, absorption and excretion – and that death quite often occurs thanks to damage to this plumbing.

A living protein, that is, a chemical molecule which enjoys primary bonding or domainial unity – rather than being a sort of sleeve containing various substitutable chemicals, which must be found by chance in the exterior world – forms a network which folds back upon itself and produces a closed membrane or open channel. This technical revolution is the point of departure not for life but for the *morphological revolution* that is habitually – and wrongly – confused with the beginning of life.

With the membrane, above all the tubular membrane, means and ends begin to be discerned in the organism. Organic networks, in which the indetermination of individuality renders ends and means indistinguishable, begin to integrate pure means into meshes, where they only participate indirectly in organic unity. Physiology in the strict sense and pure functioning begin here. The important machines that can be assembled in the organic 'factory' – machines which make use of the great masses of interrupted currents of energy and materials, machines that engage in articulations and couplings, machines which produce and respond to feedback* and constitute traps – can be understood according to the laws of classical physics and often can even be replaced by prostheses or other external technical processes.

THE RECUPERATION OF ENTROPY IN THE ORGANISM

We know that it is characteristic of the laws of crowds and the secondary laws of classical physics that they conform to the principle of the degradation of energy and an increase in entropy or disorder. The fact that disorder is inherent in the play of bonds and non-coordinated interactions is easily grasped. But the organism is capable of traversing this tendency in reverse to establish order, or at least arresting the march towards disorder. It follows that disorder derives from bonds and interactions that do not possess a unitary 'survey', the order which is itself always derived from 'surveyed' couplings and assemblages. The recuperation of entropy characteristic of the organism, which is capable of producing assemblages and couplings, is only one aspect of its absolute unity.

The organism's capacity to operate on two levels, and according to two very different processes, has not always been firmly insisted upon. Consider, for instance, the work of the renal membrane, which extracts undesirable molecules from the blood against the statistical laws of osmosis. What is normally called reverse osmosis[31] is a struggle against disorder at the microscopic level. This work, which takes place at the molecular level, is analogous to all of the intracellular labours undertaken at the level of the chemical tools adjoined to protein chains, deploying the energy accumulated in the course of chemical microcycles, for example, the high-energy phosphate bonds found in ATP molecules. At this level, the work of organisation is actualised through molecular micro-couplings.

Now consider, on the other hand, large organic cycles like those which maintain blood sugar at a constant level. As above, the result is a victory against the principles of disorder. But this time it is obtained by the macro-organisation of organs (the liver, pancreas, cerebral centres and chemical buffers) through amboception[32] or feedback* and is analogous to the assemblages of automated industry. As the cyberneticians have not failed to note, regulated automata [*automates à regulation*] arrest the degradation of energy; they are the authors of 'decisions' which maintain a given order. A cannon equipped with a regulating radar fires only at its targets and not at random; against fluctuations, a homeostat[33] maintains a certain level or assigned position.

Whether micro- or macro-coupling is at issue, though, the essential organic performance is the *assembly* of the coupled arrangement. Canalisation, the delimitation of chance through large-scale apparatuses, itself depends on a fundamental property which is already present in micro-couplings. In every case, a unitary survey must have preceded the assemblage, which is the result of a structuring activity; it is impossible to explain the canalising assemblage by the canalised flux, the dam by the flood. This is not to say that the assemblage itself can operate freely from the energetic point of view. Already at the microscopic level, organic tools make use of the chemical energy of reactions – all the more so the greater the assemblage. The construction of a dam on a river, a system of locks or a water-mill already requires the tapping of the water course itself through the temporary installations that power the machines working on the final installation. But it is not the coupling of the energetic flux and the installation that explains the installation and the coupling themselves.

'OPEN SYSTEMS'

Nothing yet has been said of the problem of morphogenesis with respect to the sentiment, remarked upon by Brillouin, Köhler, Bertalanffy and Prigione, that the organism is an 'open system in a stable state'. An 'open system' is only a model of the organism if both the *formation* and the *functioning* of the system are taken into consideration. Thus, as Needham recalls, 'today it is perfectly clear (even if many biologists continue to refuse the full significance of the point) that the organisation of living systems is not the starting point of biological research but the problem'. What is an open system in a stable state? A closed system neither gains nor loses energy and arrives at a state of equilibrium when its free energy is at its minima and its entropy at its maxima. But an open system has an entrance and an exit, like a lake, for the constant flux of energy with which it is coupled; it is nevertheless able to maintain, not an equilibrium, but a stable state, despite the variations in the incoming flux. It is thus endowed with an 'equifinality', achieving and maintaining a given state in which entropy is constant and not maximal despite the variability in the immediately preceding processes. Its stability is not the absence of movement; energy and matter are in circulation like water in a lake. While the chemical equilibria in a closed system are reversible, they are irreversible in an open system which maintains a zone of homeostasis within a flux in a constant direction. Organic processes of degradation are continually compensated for by restorations which are themselves nourished by the incoming 'nutritive' flux of energy and matter.

That organisms are open systems, and that they do not contravene the principle of the degradation of energy, is beyond question. What is even more incontestable is the fact that the functioning of an open system explains nothing about the constitution of such a system, which is necessarily a canalisation or network of ducts that always implies a formative activity. In geographical evolution, it is the circulation of water or ice itself that fashions the hydrographic network. But we no longer live in the time of Descartes when one could believe that the circulation of fluids and the deposition of sedimentation in the embryonic organism were responsible for the formation of organs. A lake only apparently possesses a stable state: the spillway wears out; the final state of a hydrographic network is an ensemble of continuous declines.

The organism, it must also be said, is destined to wear out and die. But while it is alive, and above all while it is being formed, it actively resists structural wear. The lake, by contrast, does not resist its destruction. The influx of an abnormally large amount of water translates into a larger debit on the way out, and the level remains constant, but a degradation in the spillway translates into a change in the water level without evoking a response. *The organism does two things at once*: it settles its own debts, by the catalysts that it produces by itself or indirectly procures, but it also *tries to maintain* the influx of energy and matter. It is like a lake which, miraculously, like something from Hindu myth, goes off in search of the sources that feed it.

78   As a typical example of an 'open system in a stable state', W. Köhler takes not a lake but a candle's flame. The flame, after a brief period of 'development', maintains a constant size and shape – not according to an 'equilibrium' in the strict sense, since this would mean that the potential energy of the system had been reduced to a minimum, but according to a stable system or steady state* in which it expends, moment by moment, the whole quantity of potential energy that can be drawn up into the lit wick. If the 'flame-surrounding atmosphere' pair is conceived of as a closed system, its potential energy will constantly diminish. But since the burning wick is the only site of this possible diminution, the tendency of potential energy to diminish as much as possible means that the flame contains at each moment the maximum amount of potential energy. Considering the flame in isolation could therefore lead one to believe that it contravenes the minimum total potential energy principle even though it in fact obeys it, instituting at each moment the state of maximum dissipation. Entropy only remains locally constant as a result of the maximalising principle itself. Even so, the organism only maintains its current state for as long as the flow of nutrients is available to it, and as long as it can get hold of weakly entropic (or highly organised) substances and can excrete highly entropic (or less organised) substances.

But this comparison with a flame adds absolutely nothing to the comparison with a lake. The lit wick is a spillway or a drainpipe, and schematically speaking, the candle can be assimilated with equal ease into the rhetoric of the pipe, the river or the lake. The candle is fashioned by human hands expressly to harness a constant flow of energy, and the wick, which is not repaired, is a spillway for the forecast wear; its intact portion is revealed to the extent that its reserves are diminished. Now, it is true that non-artificial equivalents exist: a piece of burning wood in a forest fire also attains a stable state; or again, to invoke something decisive for the cosmos, a normal star like the sun consumes its hydrogen not explosively but by expending its energy in a constant flow through a sort of pseudo-feedback* mechanism. But even in this latter case a

79 'membrane' or 'canalisation' effect is easy to detect since the outer layers of the star serve – as Eddington and Gamow have shown – as a heatsink thanks to which an equilibrium between the star's internal gravitational pressure and its radiation pressure is attained.[34] When this canalisation effect ceases, for reasons not yet well understood, a catastrophe takes place and the star is transformed into a nova.

THE ORGANISM AND ENERGETIC COUPLING

Considering all that can be learnt through the comparison between an organism and an open system, it is the form of the 'pipe' that has the greatest importance, in conformance with Lichtenberg's ingenious idea. From the moment that the secondary laws come to bear on the masses of physical nature, producing by chance something akin to a pipe or a spillway, the unfolding of phenomena resembles, *from the energetic point of view*, a constituted organic process. But it would be a strange thing to believe that we had thereby solved the problem of active morphogenesis. As F. Meyer[35] quite rightly remarks (and as we have also noted ourselves),[36] the coupling of 'flows of free energy', whether at 'a constant level of entropy' or even under the conditions of a 'local diminution of entropy', allows us to assert that the organism does not contravene the laws of physics, but it does not explain the organisation of the system itself. Coupling explains but itself remains unexplained. 'Even if the living being is incapable of creating negative entropy (or order) without compensation, it can nevertheless create it with compensation ... underlining the contingent comportment in relation to what thermodynamics is capable of demanding'. When it makes use of its organic reserves, the living being reorganises certain of its parts while disorganising others 'without such phenomena ever contradicting thermodynamics, but without which none of its other necessary consequences would take place'.[37] A canal, in which the locks can be both raised and lowered, does not violate the second law of thermodynamics but uses free energy, degrading it both when the locks are lowered and when they are raised. Such a structure must, however, be set up and maintained.[38] Of more importance is the fact that as the observation of its concrete comportment shows, an animal does not passively wait for flows of free energy but is responsible for them. It actively seeks out this coupling through hunting and the exploration of the milieu; it does not wait for a spontaneous exterior phenomenon in order to act but 'itself provokes this phenomenon by killing and dismembering its prey. It triggers this phenomenon, initiating its own mobilisation and efficient causation'.[39] The organism is an 'active canalisation', an energy trap that organises itself, unlike the sawmill or waterwheel, which only works because of the current of the water which does not explain its construction.

The human engineer constructs systems of pipes or energy traps and then uses the available energy to power machines, and even, when he works with his hands, makes use of organic machines. But an initial point must be reached, the point at which energetics can no longer be distinguished from formation, where 'surveying consciousness', that is, domainial unity, 'absolute surface' or 'vision-understanding', is at once the principle of energy and bonding, and the principle of formation, without use of intermediary relays or secondary bonds. As we have seen, this initial point is to be found in microphysics, in the domains of primordial individualities, and in their comportments of bonding. In such domains, there are no machines or pipes in which other subordinate individualities circulate en masse. But there is the possibility for such machines to be formed through indefinitely superposed relays, enmeshed cycles of functioning which, once established, obey the laws of thermodynamics. In their interrelations, however, they do not depend on these laws since these interrelations are founded in an individual domain where, by definition, there are no statistical laws. A virus, in its behaviour, remains the tributary of a micro-energetics which it is capable of deploying in the course of a morphogenesis that will give rise to complex organisms. At this level, functioning and formation are as indistinguishable as they are in the molecule – but from this level, two distinct paths open up, one of which leads to more or less regular masses of individuals and the other to the development of the individual organism, whose simplest schema is the formation of a tube, through the folding of a domain of individuality.

THE ORGANISM AND NUTRITION BY 'NEGATIVE ENTROPY'

We see clearly then just how misleading the popular idea, according to which the order of the organism, its 'information' in both the Aristotelian and cybernetic sense of the word, is without mystery, merely nourished by negative entropy – negentropy – and that it deploys a 'current of order'. This idea leads to the confusion of the horizontal plane of equilibria or energetic flows and the vertical plane of structuration, the canal and what is canalised. It is contrary to Frey-Wyssling's principle: *Structura omnis e structura*. The order of a structure can only give rise to a somewhat similar structure but not to *any* order whatsoever. The confusion is supported by the fact that an organism's nutrition is at once energetic and restorative of subordinated micro-structures. If the foreman of a factory received both coal and spare parts for the repair of the factory's machines on the same train, he would not be led to conclude that the 'free energy' of the coal explains the organisation of the factory. Force-feeding a goose does not promote it up the great chain of being. Business is sometimes conducted at good restaurants, but a businessman's value cannot be judged by what he eats while he is there. In order to *understand* information or organisation – for example, to produce a number of copies of a journal or to explain scientific findings to students – at least as much negentropy must be destroyed (in the form of electrical energy or alimentary chemical energy). The *sum* of information and negentropy, as L. Brillouin notes, can only remain constant. But this convertibility of the energetic order and the *extension* of information must not be confused with the convertibility of the energetic order and the structural order or the *augmentation* of information. Electricity cannot write a journal any more than a teacher can learn from food.

It is of the greatest significance that at the start of embryogenesis in mammals, during which organisation progresses very quickly through segmentation and the deployment of primordia, takes place before the implantation into the placenta, without any external flow of nutrients, and that – in the eggs of both birds and reptiles – nourishment must be accumulated significantly *earlier* that it is used by the embryo which will envelop it and fabricate circulatory mechanisms as quickly as possible.

The word 'nutrition' is often used as a legitimate and interesting metaphor to designate a structural, compositional 'profit' for the living as much as for the psychological or cultural life. In the past, one spoke of being 'nourished by literature' or 'the sciences'. We are nourished by examples, expressions which are useful to us as we work to form what it is we want to say. A polyglot assimilates the idiomatic turns of phrase of the languages that he studies; a builder receives bricks and tiles for the house he is building, but also ideas that will inspire the construction through their imitation. An animal is all the quicker to manifest its instincts as a result of incitations towards sociality; the culture of a human society is essentially a permanent possibility of 'informational nutrition'. This sort of 'vertical', morphogenetic nutrition – operating by imitation, assimilation, utilisation, resonance, incitement, 'instruction', 'good example' – represents, as it were, a 'current of order', or rather of specific orders, themselves produced only through the constructive and useful efforts of an agent dominated by a morphogenetic theme. But this current of order differs profoundly from the one with which Prigione is concerned. At the microphysical level, the two kinds of orders are still indistinct: causality, external influence, is at once energetic and 'instructive'. Elements interact through resonance and the compatibility or incompatibility of states; the influx of energy never takes place without a restructuring; kinematics and dynamics are one and the same. An atom is not like a house under construction whose builder receives the architectural plans first and the materials and sources of energy to realise later. Something of this indistinction is maintained in large molecules and viruses, but in all higher organisms the two currents of order are distinct and can no longer be taken together without risking a serious equivocation. One type of nutrition does not explain the other.

Chapter 3

Internal Reproduction

The reproduction of a form through its own activity is, alongside its active subsistence, the characteristic property of organic forms. This property is nothing other than 'verticalism' itself. In other words, it is impossible to explain the reproduction of form A as it splits in two and becomes A'-A'', for instance in cellular division, as a 'horizontal' function of A. The same holds for the reproduction of a multicellular organism as it unfolds through the emission of a germinal cell which, through division and differentiation, produces another multicellular organism similar to the first. Functioning implies that the parts connected to a machine or a system move according to their degree of freedom and without breaking their connections.

'REPRODUCTION' AND 'SELF-REPRODUCTION'

How could the functioning of a machine lead to the duplication of this machine itself? There exist machines that can reproduce numerous copies of a structure-type that is provided to them, and as we have seen, von Neumann has emphasised the fact (or the logical possibility) that this structure-type can be the same as that of the fabricating machine, 'instructions' included. *But this is only 'reproduction' and not 'self-reproduction'*. The first machine does not *become* two twinned machines in the way that a dividing cell becomes two twinned cells.

Still less can it become a double machine, a symmetrical or segmented machine, in which a unity would continue to dominate the redoubled parts that have now become its organs. So long as only the reproductive division of a virus or a gene is under consideration, we could conceivably see a 'reproduction' rather than a 'self-reproduction', see a mechanical impression then the separation of the moulding part (the template*) and the moulded part itself.[1] In all probability, this is false. It seems increasingly likely that genes, like viruses, reproduce themselves not through fission but through the active construction, induced remotely, of their own doubles. The image of a mould, borrowed from phenomena of a human scale, is not applicable to the scale of molecular architecture. How could a complex, three-dimensional construction reproduce the pattern* of its internal bonds by 'impression'? This is tantamount to claiming that the Eiffel Tower could produce another similar tower if it were placed in an enormous vat containing all of its own parts in a disassembled state. But above all and in every respect, if reproductive divisions were nothing other than instances of moulding, it would be impossible to understand developmental divisions and 'internal reproduction'.

CLEAVAGE IN DEVELOPMENT

It is impossible to understand anything about reproduction or organic development if we do not take account of the fact that total form progresses by cleavage or dominated multiplication – by segmentation and the serial or symmetrical repetition of similar parts. The first phase of the development of a multicellular organism consists of the cleavage of the initial cell, becoming two cells which are not separated, as in the division of a protozoon, but compose symmetrical parts of a unique being and which divide in their turn. In subsequent phases, the same 'internal reproduction' continues to play a central role, where no longer cells but entire primordia divide. Comparable segments – metameres and somites – appear in the embryo of the earthworm, and traces of this segmentary formation, however dissimulated by subsequent development, can easily be seen in the adult organism. 'Internal reproduction' can take a radial form, as it does in starfish, sea urchins and medusae. Development can also resemble, in animals as often as plants, various forms of bifurcation, animal colonies resembling trees or a bed of flowers.

The fact that we have two hands, two eyes and two similar kidneys while being a single being is a decisive challenge to those mathematicians and physicists who claim to have explained reproduction – we should instead say '*self-reproduction*' – through the functioning of a reproducing machine. When a cell divides, whether in an egg or a multicellular organism, where, how and of what is there functioning? The different phases of functioning must by definition be able to be correlated 'one to one'. The group $a' + a''$ cannot be correlated with an a precisely because both a' and a'' resemble a trait for trait. The fact that – *in* the set $a'\,a''$ – both a' and a'' can be isomorphically correlated 'one to one' with a excludes the possibility of a 'one to one' isomorphic correlation of this set with a. The derivation cannot be the result of functioning. In turn, a'' cannot be the result of the functioning of a', or vice versa, since both a' and a'' simultaneously appear in cell a which, in the first phases of division, is a unique form with a plane of symmetry. It is equally absurd to assert either that our left hand results from the functioning of our right hand or that our two hands are the result of the functioning of a unique structure contained in the egg from which they arise. It is equally absurd to assert that one of a pair of twins is the result of the other's functioning, or that they are both the result of the functioning of a unique, primitive egg. Internal reproduction manifests, to a greater extent than numerical reproduction, the 'vertical' action of a formal theme. The serial arrangement of the rudiments of an organ – quite evident in the fin of a skate or the embryonic bud of a human hand – appears as an active principle of structuration, 'an opportunity of making available in a simple form the building material required for producing more complicated shapes of the body'.[2] This simple force, acquired through segmentation, is only retained to the extent that the adult organism can accommodate it. Segmentation and bilateral, radial or bifurcated symmetry are subordinated to a morphogenetic theme whose action is not explicable in mechanical terms. A vertical hierarchy of formal themes is established in which initially similar parts – which are dominated while nevertheless maintaining a semi-individuality – can then differentiate themselves within a unitary domain. This, rather than the multiplication of form, makes progress progress.

INTERNAL REPRODUCTION AND EQUIPOTENTIALITY

The contemporary neo-vitalism of Driesch arose precisely as a result of the shock of encountering 'internal reproduction': half of a sea urchin, just one of the two cells resulting from the first division of the egg, can produce an entire embryo. But this is not what struck Driesch. He describes this phenomenon in a general but nonetheless quite correct way as 'equipotentiality' and shows, no less correctly, that the equipotentiality of both part and whole rules out any possibility of a mechanist explanation.[3] The argument by which he establishes this is effectively the same as that which explains the auto-division of a form through internal reproduction and not the functioning of this form. Take the rectangle ab (figure 3.1), which represents normal development – for instance, that of a sea urchin egg in which no scientist has intervened.

Each part of the rectangle has, let's say, a prospective 'destiny', for instance, the point x. This point or its region will become, let's say, nervous tissue. But if, by isolating $a'b'$ or $a''b''$, normal development and a complete individual can still be obtained, x will have had to become something other than what it would have in ab, and, more than possessing a prospective signification, it must have substantially vaster 'prospective potentiality'. Each part must therefore possess a relation to a 'factor of totality', E.

Driesch's argument amounts to saying that there cannot be a 'one to one' relation between phase I and phase II of development. In other words, if

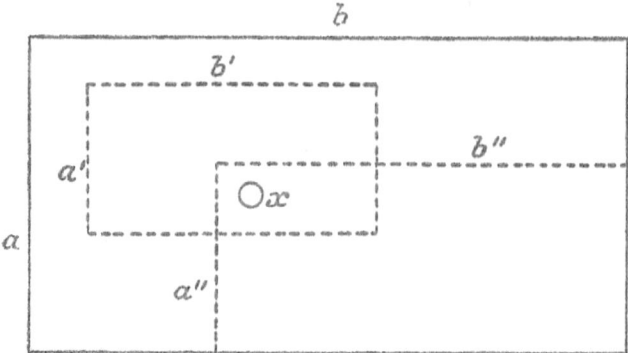

Fig. 3.1

89 one part can remake the whole, or if the whole makes its own parts not by cutting them out but by reproducing itself, then functioning cannot be made to account for internal reproduction. And despite the clumsiness of his terminology, Driesch is not wrong to make reference to mysterious entities or to define equipotentiality in the most general terms since it appears not only in the first stages of development but in the mode of development of respective embryonic areas, in regeneration, in the accidental fusion of the outlines of two paired organs – and this is not yet to speak of the psycho-biological phenomena of perception, memory, habit and learning, which manifest the equipotentiality of the brain or cerebral zones.

In the particular case of the sea urchin egg, though, the facts appear to remain open to the possibility of a simpler explanation which would seem to dispense with the need to invoke a factor of totality E or an entelechy. It would appear to suffice to admit that the first divisions of the egg produce two, four or eight primary cells and that these divisions are in fact 'reproductive divisions', no different from the numerical reproductive divisions of a protozoon. There is nothing striking about the fact that two progeny amoebae act like their progenitor – why then be struck when each of the two or four primary cells of a sea urchin's egg can become a sea urchin as well as a whole egg? The organism's capacity for reproduction seemingly allows us to dispense with recourse to the mysterious property of equipotentiality and furthermore explains the structural progress and augmentation of complexity produced by development. After numerical multiplication, the adjoined individuals occupy new positions relative to each other and establish new interrelations, which lead to differentiations of role in keeping with this new structure. As such, it is not surprising that a society of eight or sixteen individuals has properties that differ from an isolated individual, even if these eight or sixteen individuals were initially similar to each other. If it is agreed that the spatial repetition of a form requires no entelechy, and that new properties derived from new relations can emerge naturally from a society of individuals, development no longer possesses any mystery. The mystery of equipotentiality is dissipated along with that of epigenesis.[4]

Unfortunately, this 'explanation' is worthless since, as we have seen, reproduction itself presupposes equipotentiality; it is only equipotentiality in another form, and the two properties are only different manifestations of the 'verticalism' or thematism of organic forms. In the development of the sea urchin and of any other regulating organism,[5] the first divisions can pass as reproductive divisions, where the products simply remain adjoined. But even if we presuppose that it is solely numerical in the first instance, numerical reproduction must, progressively or abruptly, become internal reproduction, and the progeny-cells must become parts of a being after having each been one. What is essential is precisely this possibility of a hesitation between two types of reproduction.

INTERNAL REPRODUCTION AND THE AUGMENTATION OF COMPLEXITY

Correlatively, and contrary to Woodger's contention, the simple repetition of an object in space, followed by the adjunction or spatial assembly of the resulting object, is completely insufficient to explain the augmentation of complexity in a system. It is undeniable that a mass *qua* mass obeys different laws than those which apply to the objects that compose it. The science of oceanography is completely different from the chemistry of water or salt. But to invoke 'different laws' is not also to invoke 'augmented complexity'. The ocean is not a being whose organs would be molecules of water or salt, and the structure of a wave is certainly simpler than the structure of a molecule of water. A chunk of sandstone is only aggregated sand, and as a rock it is not a more complex being than each of the grains of sand or molecules of silicate that it contains. These molecules, furthermore, are not really differentiated according to their place in the whole. We will see that there is a whole possible sociology of organic forms and their development, but only on the condition that the word 'society' is taken in its true sense and not understood as a simple juxtaposition of individuals. A society in general always implies that the individuals that compose it obey a series of coordinating themes in every respect and that they know to play their 'roles' according to diverse situations-stimuli, 'roles' which do not arise automatically as an effect does from a cause, from the simple spatial situation of the individual in a social ensemble. The mystery of differentiation can be dissipated by considering it to be the effect of situational differences produced by equal divisions. These differences are stimuli and not causes.

THE CASE OF *VOLVOX*

Let's consider the case of *Volvox* and similar colonial forms in which every form of transition can be found, from poorly integrated cellular colonies to those so well unified that they resemble a unique individual.[6]

> The *Volvox* is composed of around two thousand green cells possessing two flagellae, arranged in a slightly elongated hollow sphere. The cells are quite similar to one another, though a certain differentiation and division of work is apparent. The elongated axis of the sphere or ellipsoid determines a polarity, around which *Volvox* turns, and the cells of the uppermost hemisphere are the largest and most vivid in colour. Strictly speaking, this difference could be explained, as Woodger does, by the fact that they are situated at the top and receive the most direct light. But it is certainly not this situation alone that explains the curious mode of reproduction of the colony. The reproductive cells belong exclusively to the southern hemisphere. The swollen reproductive cells divide, forming a sort of pocket facing the interior of the sphere (figure 3.2). The daughter colony then completely detaches itself and swims into the interior of the mother colony, to be freed only when the latter dies. But the remarkable fact is that at the moment the daughter colony detaches itself, it turns itself inside out like a glove, thus directing its flagellae towards the outside.

We have here a striking example of the fact that a 'role' does not result from a single circumstance but that a given role, possessing its own thematic consistency, *searches for an appropriate situation*. This is far from an isolated example – sponges, for instance, engage in the same involution. In the great majority of cases, one observes differentiations of role and structure in constituent cells which *seem* to result from their place in the whole and, at the same time, morphogenetic migrations towards an *appropriate situation* for the role and for the whole constituted structure.

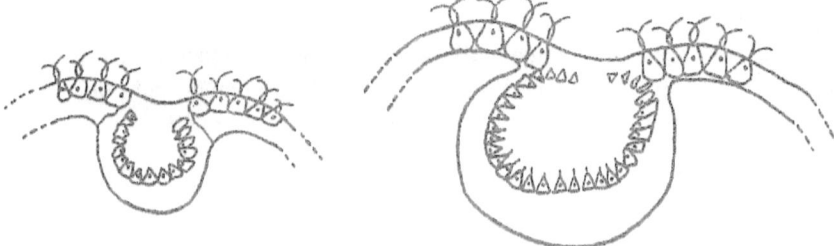

Fig. 3.2

We know that the cells of a sponge, having been dissociated and dispersed, are not slow in re-aggregating and, at around four days, are reformed into a sponge-type with its many functions. The cells maintain their roles and differentiations despite their diffusion, and they adopt the same places in the new sponge according to this role. And yet, it is probably the case that in the wake of this operation, certain cells might sometimes change their type.

The case of the hydra is even more revealing. Since Trembley, we have known that this animal possesses remarkable properties of regeneration. With the exception of the tentacles, any fragment of a hydra can, without absorbing any nutrients, reconstitute the missing parts by rearranging those that remain. This operation implies that the cells which formerly undertook a given function in a certain location are modified and adopt a new function in the regenerated organism. Nevertheless, the cells of the endoderm and the exoderm are not interchangeable. A mass of endodermic cells is incapable of remaining united; a mass of ectodermic cells can only form an elongated sphere which is not differentiated; both kinds of cells are required in order for the fragment to regenerate. In a hydra turned inside out like a glove, regulation can only take place by cellular migration and not through differentiation in place: once determined, the 'role' determines the location. We can therefore conclude that even in those cases where location alone seems to determine role and differentiation, 'determination' in the biological sense of the word is at issue – that is, the fixation of an entire 'destiny' according to stimulus rather than any kind of determinism through cause and effect.

The case of *Volvox* is also of interest in a further respect. The cells possessing flagellae contribute movement to the whole of the colony, whether they are relatively autonomous individuals in a society or the supports of a unique theme of movement. When *Volvox* swims towards a light source, this movement of the whole is in reality obtained through the individual responses of its constituent cells: the flagellae on the unilluminated side flutter in an effective, dissymmetrical fashion, while the illuminated flagellae only engage in 'neutral' movements. In contrast, the movement of rotation in either direction depends on a kind of wave of movement which dominates the cell's individual activities.

COLONIAL AMOEBAE

The amoebic colony *Dictyostelium*, recently studied by Bonner,[7] clearly demonstrates the passage from reproductive division to 'totalitarian' differentiation. Its life cycle begins with the amoebae, themselves born from spores. These amoebae initially lead an independent, individual life; they move around like protozoa and actively multiply. But when a critical density is achieved, they seem to become attracted to each other.

Some begin to assemble first, and then in converging currents, a mass that includes billions of amoebae forms (figures 3.3a–b). This mass then begins a general movement, one that in fact depends on the amoeboid displacements of its constituent cells, the amoebae of the head leaving behind them a trace

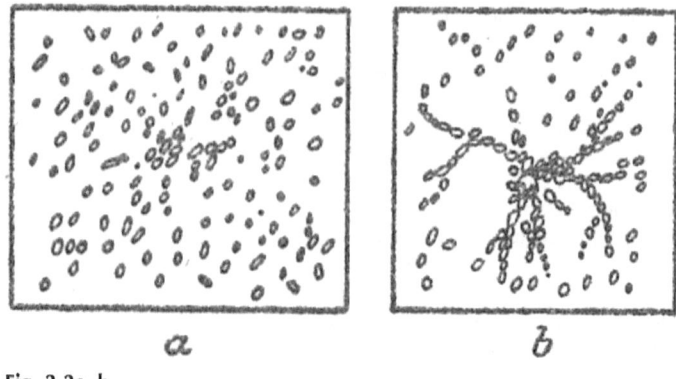

Fig. 3.3a–b

that seems to guide the others. This mass then differentiates itself in a more complex fashion: it splits vertically and forms a sort of mushroom with a thin, elongated foot and a rounded head, within which a certain number of cells transform themselves into spores. The elongated foot is also formed through the differentiation of its constituent amoebae, which close up, vacuolising into a cellulose sheath (figures 3.3c–d).

These morphogenetic movements of assembly, migration and fructification are probably triggered and guided by special stimuli. These stimuli have been studied (by Pott, Runyon and Bonner), above all in the case of assembly, in which, it can be shown, chemical substances that act not as causes but as signals-stimuli – rather than either an electrical field or any kind of radiation – are at work. But the problem is already much more difficult for the migration of the whole colony. It seems that the cells of the head emit greater amounts of the guiding substance, thereby constituting a gradient. But it is difficult to see how this can be the initial trigger for the head-tail polarity and

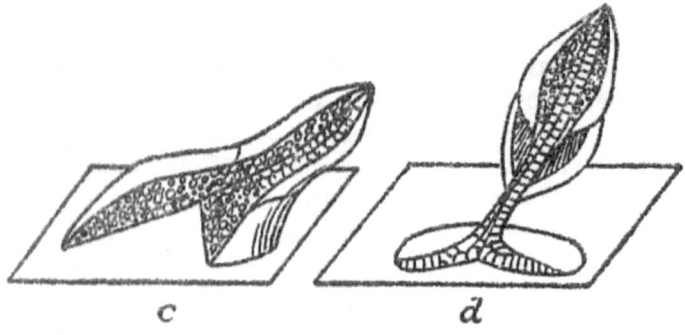

Fig. 3.3c–d

the unequal spatial distribution of substance, which is produced even when one attempts to prevent any dissymmetry by experimentally agitating the colony. It is even more difficult again to explain the complete morphogenesis of the spore bearers as a sum of the individual behaviours of its constituent amoebae. It is clearly the case, not only that each individual amoeba must be 'competent' in order to respond to diverse stimuli in diverse situations but that it must virtually be the bearer of the whole 'colonial' theme since an entire colony can be born from this single amoeba.

It is impossible in any case to claim that identical amoebic reproductive divisions (during the pre-colonial phase) explain the formation of the colony, its structural development, its augmentation in complexity, its equipotential character and its possibilities for regeneration (which manifest themselves whenever the cluster or migrant colony is divided) since there is a reproduction of the colony as such, one that goes through a phase of growth – the reproduction of individual amoebae – before a phase of differentiation without growth, taking place through morphogenetic movements of differentiation, begins.

The same hesitation between 'being an individual' and 'being the organ of an individual' is found throughout the organic domain. It is found, as we have seen, in the flagellae cells of *Volvox*, which can execute individual movements and the colonial movements of the whole equally well. It is found yet again in colonial bacteria or myxobacteria, discovered and described in 1892 by R. Thaxter, which have a life cycle similar to that of a colony of amoebae. It is perhaps also found, if the work of certain Soviet biologists is to be believed, in the life cycles of certain bacteria, considered as colonies of molecules, which undergo a dispersed molecular phase. It is also found, finally, in the cells of metazoa in development, or indeed in any organism whatsoever. The relations between internal reproduction and numerical reproduction in morphogenesis can only be understood by invoking a non-mechanical model, by thinking of an individualised melodic theme which can both be integrally repeated and distribute itself in variations through which the initial, repeated theme serves as its own 'development' (in the musical sense of the term).

THE ATTENUATION OF PERSONAL PRONOUNS

Once again, our use of the 'psychological model' must not lead us to be duped by a conventional psychology that imitates the simplifications of language. The 'linguistic' character of Aristotle's philosophy is often denounced, along with the 'Indo-European' character of his substantialism, which treats the subject of the verb as if it were a metaphysical substance. The grammatical list of pronouns in modern languages, however, imposes an equally insidious 'invisible postulate'. *I, You, He, We* and *They* appear to be separated by an unbroachable abyss.⁸ Consequently, the principles of individuation and their possible relations appear to us to necessarily obey the linguistic and social laws of the relations between completely constituted and numerically distinct individuals in social space.

All of biology and psychology nevertheless show that this postulate must be rejected and that personal pronouns must be 'attenuated'. 'I' am not really a pure I.⁹ When *I remember*, when *He remembers me*, or when *I dream*, a mnemic *I*, an *I* at once other and myself, merges with the current I. As psychoanalysis has clearly shown, a whole intermediary series takes place between *I* and *He*, supporting the complexity of my psychic architecture. 'I' am made of all these other *I*s that I have produced through a kind of cellular division of internal and dominated reproduction. I am as much a psychological as a biological colony. We say, 'I'm shivering' and 'I breathe' even though our muscles are obeying almost completely autonomous rhythms. 'Its tongue moves, or it moves its tongue', says A. Gesell of the human embryo at the uncertain stage in which movements that much later will be voluntary have not yet become completely coordinated. Doubtless in this precise case, we can admit that the passage from one stage to another can, in principle, be precisely marked by a mechanical and spatial phenomenon, namely, the establishment or first functioning of a nervous connection. But there are numerous analogous cases in which such a mark is impossible to discern. At what point could it be said of a sea urchin, '*There is* a right and left side' or '*There are* two similar adjoined cells'? Do amoebae move in groups, or does the colony itself migrate? Could *Volvox* say, '*I* turn myself around', or could its constituent cells say instead, 'I make my flagellae move in keeping with what my neighbours are doing'?

Chapter 4

The Fragmentation and Socialisation of Development

Through internal reproduction and the 'distribution' of the theme, fragmentation and multiplicity are introduced into development, which loses its unitary aspect.

As any familiarity with treatises on embryology will reveal, the most extensive chapters are those devoted to experimental research on interconnections of development, on fragmentary sequences and on the stimuli which determine this or that partial development. And yet there is no absolute opposition between the unitary and fragmented aspects of development. Vertical thematism is no less evident in fragmented development and the play of stimulus and response than it is in unitary development.

INTERCOMMUNICATION IN DEVELOPMENT

Even if there is in fact anything 'step by step' in development, the stimuli that play the role of 'causes' and seem to determine the sequence of phases or interaction of parts are not in fact causes, in the sense that a falling domino is the cause of the next domino's fall. Stimuli are signals, evocateurs. They provoke responses in the competent stimulated parts; they provoke behaviour or formations whose degree of complexity greatly exceeds that of the signal stimuli. As soon as the fragmentation of a specific theme becomes irreversible – when the indetermination of individuality which allows for the fusion or doubling of primordia is no longer in play, as it is in the first stage of the development of the egg – an intercommunication between parts must be produced. 'Social' relations, in the strict sense of the word, which imply media of communication, active signification and 'language', must be established in order for harmony to be maintained.

Fragmented development is always secondary in relation to total thematic development. The sequencing of signals and responses is a kind of 'montage', a surrogate automatism. Despite appearing to constitute causality *a tergo*, the fact that these sequences normally lead to a valid result and not just any result at all demonstrates the secondary character of fragmented development. But the intercommunications and sequences play no less of a decisive role in morphogenesis, in which veritable 'social' instincts are at work, making the parts of the organism adjust to one another in the same way that ordinary social instincts make the individuals of a colony adjust to one another.

The most recent progress in the science of morphogenesis is, it can be said, to be found in the work not of embryologists but of psychologists of instinct – Tinbergen and his school, and Lorenz and von Frisch, who have studied the role and nature of signal stimuli in instinct and in the social coordination of organisms. Detailed study of the interconnections between releasers* and motor melodies in response – in the sexual behaviour of the stickleback, for example, or in the family life of the seagull – is a direct contribution to the knowledge of morphogenesis.[1] As Tinbergen notes (and the bacterial or amoebic colonies studied by Thaxter and Bonner illustrate this thesis perfectly),

> many animal communities depend on the functioning of remarkably few and simple relations. Whether a community differentiates from a simple organ-relationship, or is constructed by two independent individuals joining into an organisation, the relations between the individuals, based on the releaser-system, begin to function as soon as they are needed, or even before [. . .] When studying the way in which a community is organised, one is often struck by the many parallels that can be drawn between it and an individual. Both are composed of constituent parts; the individual is composed of organs, the community of individuals. In both, there is division of labour between the component parts. In both, the parts co-operate for the benefit of the whole, and through it for their own benefit. The constituent partners give and receive.[2]

ORGANIC AND SOCIOLOGICAL COMMUNITIES

There are obvious differences between organic communities or animal societies and human society. Relations of stimulus and response, intercommunication and the cultural or political performance of 'roles' do not at all have the same origin in human society, where they are clearly not the result of a potential or primary theme of which they would only be a fragment. Human language, transmitted by education and psychological imitation, is profoundly different from the 'language' of bees, which has not been individually learnt. Human beings, even in so-called primitive societies, are both conditioned by a non-hereditary culture and calculators, players or strategists in most of their action. These differences have always lent sociological theories founded on biology a superficial air, just as they do to the inverse and ingenious, if somewhat forced, attempt by Dupréel to found biology on sociology.[3] And yet, if we allow that the actual appearance of phenomena can be considered without reference to differences of origin and nature, then a rapprochement can be established between organic communities and human society. A human society only exists on the basis of a potential of stimuli and typical response, occasions and typical roles maintained in cultural capital. Social behaviour is never in fact a pure improvisation on the part of the individual. The structure of institutions results in actions in which utilitarian calculus is ritualised and conventionalised.[4]

> This is why, for lack of a uniform code of stimulus and response, the same tragic misunderstandings are found between different human cultures as there are between different species. If, in an instinctive or cultural rite, a submissive attitude is taken instead to be an attitude of menace or of complete defeat, the same unfortunate results accrue to both humans and animals. Lorenz cites the case of the unfortunate turkey who offers his throat to the attacks of a peacock, adopting an instinctive attitude of submission normally destined to disarm its adversary, but one that the peacock does not understand: 'Encounters between a turkey and a peacock are always tragic, and they happen frequently, since the two species, being related, make expressive movements that are somewhat analogous to understand. When the turkey adopts a submissive posture (which would completely disarm others of its kind in similar battles), something terrible happens: the peacock does not understand this posture, and it continues to set itself upon an adversary which obstinately refuses to budge'.[5] Analogous adventures occurred when the American Indians encountered the Spanish and when European explorers encountered African tribes: 'early explorers among the Kikuyu of East Africa took a mark of hostility for an honorific welcome. . . . Amicable relations adopted a completely different appearance among these people'.[6] By mixing in the same hive Italian and Carniole bees, whose 'languages' are somewhat different – the Italians dance less quickly, indicating a given distance by dancing in a straight line with fewer vibrations per minute than the Carniole bees – von Frisch provoked confusion and errors.[7] It is quite likely that difficulties in hybridisation and the infertility of hybrids also arise due to a lack of 'social' adjustment between stimuli and responses lacking an identical instinctive code.

If we take all the facts of morphogenesis for all individual and social formations together – as we are justified in doing – and if we can give the name 'potential' to the vertical themes which assure formation, then we can say that the potential of a species is a 'collective' whole in a 'dispersed' or 'social' state in the sense that Whitehead uses the word. The body of an animal 'is a society involving a vast number of occasions, spatially and temporally coordinated'.[8] A species, with its male, female or neutral types of organisms, its castes, its individual or social types and its forms and behaviours embryonic, larval or adult is essentially a bundle [*un lot*] of themes which can be evoked, in a multitude of circumstances, by inter-stimulation or by the stimulation of a situation-type of an habitual milieu. It is, despite its difference in origins, a kind of biological 'culture' analogous to the culture or ensemble of 'roles' and social attitudes in human society.

> An even more precise comparison would be that of the morphogenesis of all the forms of a living species and the morphogenesis of a real but 'ideal' human society such as appears in the

103 reveries of 'hedonist' utopians like Rabelais or Fourier. A multicellular organism, in all of its organic or colonial forms, is a sort of Abbey of Thélème[9] or phalanstery where the instincts of constituent individuals adapt themselves, without localised government or external constraint, to the roles distributed by the single game of stimulus and response which, however capricious in appearance, is ruled by harmonious and 'providential' instincts. Properly understood, this rapprochement is in fact considerably less forced than the completely biological model that inspired Rabelais and Fourier, who saw, or believed they had seen, a spontaneous instinctive register and natural providence in the vital order that they wanted to transpose into the human register. Fourier would have been overjoyed to read Tinbergen's phrase describing the 'remarkably few and simple relations'[10] required for a communitarian organisation: he would have found in it a confirmation of his theory of the 'fundamental passions'. Applied to human beings, the sociology of a Rabelais or a Fourier is puerile, but it constitutes a very good representation of an organic 'sociology'. In a healthy organism, cells or tissues do what they have to do without any constraint, instinctively responding to the incitations arising from neighbouring cells or from the nervous or hormonal systems. Though 'cellular anarchy' is sometimes metaphorically invoked, nothing like an anarchy-revolt is at issue but instead the errors of formative instincts, betrayed by accidentally modified stimuli.

It is true that extremely intimidating moments of social constraint have been witnessed: for instance, the massacre of males by worker bees or the brutal expulsion of reproductive ants from *Atta* colonies. These constraints, however, never have a 'governmental' origin. Specialists in insect society have recently discovered the role played by spontaneous leaders in constructive activities, where more-active individuals function as a 'centre of excitation'. Certain ants, for instance, play this role 'not by giving instructions to other ants, but by responding before others to the stimulus of 'work to do'.[11]

104 It is impossible not to think of Fourier's description of a pleasant development here. To cite D. W. Morley once more,

> When twenty or thirty ant leaders get to work, they immediately excite the need to eat or go out in search of provisions, the instinct to repair a damaged part of the nest or to construct new rooms for the brood in others. . . . The work executed by each ant changes when new stimuli comes into play. . . . When work underway is almost finished or when an ant leader has worked for a long time and grown tired, the attractive power which had others in its grasp diminishes. The workers become more sensitive to the attractive power of other tasks. . . . Thus, through attraction and counter-attraction, the complex series of tasks is commenced, continued and successfully completed.[12]

Morley himself also notes the case in which, with respect to communications between individuals through a contact of antennae, odours or stridulations, the messages of an ant leader excite something akin to an habitual stimulus: 'I am quite occupied in doing this' seems to be transformed into a directive: 'Come here and do this'. 'It even happens that the excitable ant takes another and leads it by force to a spot where food has been discovered'.[13] But this is very rare, and in normal behaviour, constraint – which, in human politics, always betrays a lack of ingenuity according to Fourier – plays a very small part.

In the course of studying the building behaviour of the wasps *Polistes*, E. P. Deleurance highlights another characteristic fragmentation of the general theme of construction, one which does not essentially proceed by the division of labour and the inter-individual play of stimulus and response but which depends on internal factors and which once again evokes Fourier and his 'alternating, contrasting'[14] passion, even though precise observation has not been able to demonstrate a mathematical regularity in the periodic cycles of instinctive drives.[15]

> The wasp's nest is constructed on a stalk that supports a honeycomb-like foundation. The wasps work on it in cycles, according to the periodic manifestation of an internal impulse which repeats itself, while diminishing, a number of times each day. The parts of the nest are always successively dealt with by each wasp according to a typical order: they begin by working on the stalk, then move on to begin a new cell, and then move up to a pre-existing cell. The wasp's behaviour appears to be composed of independent and pre-determined segments, and the construction of the whole is realised through the simple repetition of the cyclical theme. The wasp obstinately pursues the theme without being directly guided by what it has already realised. Even though the repair work depends on an on-going theme, uninformed observers often believe they are witnessing a voluntary, organised effort to repair the damaged nest such that the repair work depends on an on-going theme.

The contrast between the adjustment of roles by social stimuli in ants, and the social construction through repetition of an individual theme of work in wasps, is more apparent than real. It is always a question of a fragmentation of specific potential, of a division of labour in morphogenesis which, by inter-adjustment or the repetition of disassociated themes, nevertheless succeeds, in however bumbling a manner, in producing an harmonious formation, mimicking a true division of labour or a true calculated repartition. The wasp is not a pure psychic automaton: it broadly takes its perceptions into account, at least in order to recognise the stalk, the honeycomb and the presence of an active brood. For its part, the ant certainly manifests the periodic internal rhythms which make it more or less alert to stimuli. What is key is that external stimuli, like internal stimuli or rhythms, are rigorously specific. Everything fits together because each of the partial themes is derived from a dominant theme: they are the 'phases' of a unitary theme and thereby simply reconstitute a primitive unity. The factors which would bring about a utopia in human society – because, save from the perspective of a mystical providentialism, it does not possess a specific, normal political form, human beings must engage in a hazardous invention of social forms – do not do so for organic societies, where an organic 'providence' (however this is to be interpreted philosophically, and whether or not it arises through natural selection) is on the contrary a patent fact.

This provisional conflation of the morphogenesis of a multicellular organism and that of an animal society must not lead us to forget the significantly greater precision displayed in individual morphogenesis, where partial developments are modified and controlled with the precision of a watch, and where nothing seems to be left to chance encounter. The formation of an individual, above all an animal, takes place at a rigorously defined time and in a rigorously defined way. The least structural details are specific, and even individual characteristics are genetically 'determined', as the resemblance between twins demonstrates. Despite its specificity, a colony or an animal society is not formed according to precise timing, and it does not lead to a structure which would be able to be superimposed upon that society's twin. The diverse castes, complex forms and behaviours of an anthill all arise from a single initial queen – just as the differentiated forms of a metazoan all arise from the egg, but with a great deal less precision. The 'social' character of individual morphogenesis is also revealed, no less well if more crudely, by social morphogenesis. The fragmentation of potential and the processes of adjustment by inter-communication, stimuli and response, by the evocation of autonomous sub-themes, are the same in both cases.

If individual formation is more rigorous, it is because the conditions of embryonic development are rigorously standardised. Stimuli do not come from an exterior milieu completely by chance or due to the vagaries of geography but from another organic locale. Artificial intervention is required in order to put an embryo in the awkward and unforeseen situations that societies of ants or bees confront at every instant in the external world. Gastrulation resembles a cellular migration, but it is a migration in which the cells, unlike those of birds or fish, are certain to find what they are 'searching' for and will arrive in the domain that they 'expected'. The encounter between two sexual partners is always an adventure, to the extent that in French, the word 'adventure' has acquired the particular sense of an 'amorous adventure'. This is not the case, or is so to a lesser degree, in the encounter between germinal cells; even less is it so in the formation of these cells. The internal life of the organism is innocent, or all but innocent, of the dangers of love, war and the hunt, the three great sources of 'adventure' for the living. There are still hazards in the encounter between a spermatozoid and an egg, once internal fertilisation has taken place, but these hazards are, all the same, more 'canalised' than they are for the encounter between male and female which was their precondition.

What is nonetheless key is the fact that these encounters are of the same nature, and this is why every possible transition exists between rigorous morphogeneses, which have the appearance of clockwork, and 'open', 'adventurous' morphogeneses. A plant, which must model itself on its physical milieu, already has a less specific and precise form than an animal. A plant is much more clearly 'colonial', and its formation more closely resembles an historical and accidental development.

The error would be to make of the exterior milieu a cause rather than a stimulus that provokes specific responses. The advantage of interrelating individual and social formation is that it guards against this error and allows for a concept of development which can be applied to both the most open of forms and the most closed; to both the indirect and the direct products of life; to both the forms in which vital themes have had to compromise with the accidental dispositions of the physical world and those forms in which life is guaranteed a zone of provisional shelter and a total mastery of events.

Chapter 5

Signal Stimuli

Development does not have 'causes'. In 'socialised' development, what is taken as a cause is in fact a signal stimulus. The notion of signal stimulus was elaborated by E. S. Russell and Tinbergen in the context of an experimental study of instinct. When the partridge emits a 'warning cry', its young go to ground and remain absolutely motionless; chicks two or three hours old huddle down just as well as those of three or four weeks.[1] Partridges brooded over by a hen do not huddle down even when the mother hen lets out a squawk of terror, but if a specific 'cluck' is imitated, they will nestle down immediately. When the female stickleback enters the nest while already responding to the specific stimulus of dancing zig-zag with a male, it will only lay its eggs if the male strikes its snout, rapidly and rhythmically, at the base of her tail. In its aggressive phase, a male red-breasted robin will attack a tuft of red feathers. When a young fox becomes old enough to eat meat, it is indifferent to a piece of rabbit flesh but suddenly becomes animated if a piece of fur remains attached to it.[2]

Human hunters, fishermen and farmers have for centuries made use of imitations of signal stimuli in order to corner and trap their prey, just as animal predators did for millions of years before them. Signal stimuli are, as we have seen, the means of ordinary coordination in a colony or organic society. It seems inevitable, consequently, that – given the impossibility of establishing a precise frontier between colony and unitary organism, and at least as a first approximation – we also interpret the means of coordination deployed by both the adult organism and the organism in development as signal stimuli. The substance emitted by the constituent amoebae of the *Dictyostelium* that guides amoebic concentration or migration is clearly indiscernible from a hormone or an 'organisin' of development.[3]

_{So long as a king and queen are present in a termite colony, workers and soldiers remain sterile. But should they die, they will be immediately replaced by secondary reproducers whose development had previously been inhibited. The saliva constantly exchanged between termites is probably the vehicle for the inhibiting substance; it is quite likely that the effect of 'social hormones' of the same kind are what determine their castes. As the recent experiments of M. Lüscher have shown,[4] other signal stimuli, transmitted not through saliva but through the contact of antennae, are at work in the maintenance of secondary reproducers – should the antennae}

of secondary reproducers touch a neighbouring colony endowed with a king and queen through the bars of a cage, they will be killed by their own colony.

When, in experiments on parthenogenesis, the development of an unfertilised egg is artificially triggered by pricking it with a needle dipped in blood plasma, the signal stimuli is used as a 'biological ploy'. It should not be concluded, as was done initially, that a cause has been discovered. In the same way, what are known as 'organising' substances – primary or secondary inductors – act as stimuli and not causes. Embryonic tissue is 'tricked' by an artificial stimulus, or artificially displaced when, by the drawing of a substance belonging to what is known as an organising zone into a new location, a neural canal and a secondary embryo are formed; without this false 'signal', the tissue would only have developed into the standard epidermis. The epidermis can be tricked in the same way by transplanting the circular tympanic cartilage somewhere other than where it would normally form an external ear, thereby inducing it to form a tympan. *In short, the entirety of experimental embryology, which remains voluntarily committed to a 'classical' science based on the notions of causality and determinism, is at root an ensemble of experiments concerned with signal stimuli and their ploys.* These experiments are analogous to those which aim, in the psychology of animal instinct, to discover what in instinct is specific stimulus and what depends instead on the autonomous unfolding of acts or on improvised regulation.

AN ATTEMPT AT THE CLASSIFICATION OF 'EFFICACITY'

In order to better understand the nature of signal stimuli and their difference from causes on the one hand, and signs on the other, it is worth reviewing all forms of 'efficacity', to use the vaguest possible word.

A. If I inflate a bicycle tyre, its shape and its internal pressure are the obvious and direct effects of the pumping. In classical physics, cause acts through pressure and accumulation, producing proportional effects. It allows for measurement and the establishing of laws-functions. For example, the movement of the needle of an air-pressure gauge is the direct and proportional effect of pressure. This type of cause, which always puts into play a multitude of elements, taken en masse, is not fundamental, as everyone today recognises.[5]

B. If I set up some dominoes in a line and at the right distance from each other, the fall of the first, initiated by me, leads to the successive fall of all of the others. The effect is not proportional to the cause. And yet the fall of a single domino, initiated by the slightest impetus, is continued in it, and the falling of the whole row clearly depends on their existing arrangement. On the one hand, the role of 'potential' can already be discerned in effects of this kind as can, on the other hand, the role of the disposition or relations within the system. And yet, this is no longer a case of pure causality without any signalling or information: the fall 'is communicated', but the dominoes do not 'communicate' between themselves. The reason for the effect is visible in space (although, in all rigour, the potential of the force of gravity is not 'observable').

C. A trap is set, or a gun is loaded. Any mechanical stimulus on the trigger whatsoever can activate the spring which will, in turn, set off the trap or the weapon, according to its structure. In all triggering of this kind, the indifference of the form of the effect to the form of the cause is striking; the trap or the gun functions according to the mechanical or chemical potential of the spring or the powder, and according to the arrangement of its pieces. It is also the case that the cause is often described as a 'stimulus'. It is not, however, a signal. It puts a potential into play, but it is not confined to the gestures of a pantograph.[6] But this invisible potential is nevertheless legible through the knowledge of physical or chemical laws; it does not itself create form, but it moves an existing form in space.

D. Stimuli-triggers can be deployed as stimulus-keys, another part of the trap being deployed as a 'keyhole'. By these means, the stimulus *imitates* the action of a signal stimulus. It can be said, somewhat humorously, that the trap or any apparatus homologous to it 'recognises' or 'perceives' the trigger and that the trigger conveys information to the trap. But this is obviously no more than a metaphor.

It is perfectly clear that none of these four modes of causal efficacy are suitable to interpret biological efficacy itself such as it appears in formative or instinctive interactions. They are all found, of course, in the functioning of the organism, even the effects of proportional pressure, the effects of mechanical traps in particular. Both the leaves of the Venus flytrap and nematocysts[7] presumably function like a gun or a spring-loaded trap, and even if the system is different, it is of the same order. There also probably exist key effects.[8] Cyberneticians, on the one hand, and mechanist biologists and psychologists on the other, have tried to interpret organic behaviour in its totality on the basis of these kinds of effects. Automata equipped with organs of 'perception' or 'information' – which in reality are, with or without a 'turnkey transformer', only 'lock-traps'[9] like the reading machines found in the work of Pitts and McCulloch[10] – are given as authentic models of the organism. Pavlov's attempts to interpret conditioning stimuli as the triggers of cerebral analysers, and not as truly perceived signals in the psychological sense of the term, go in the same direction.

It suffices to consider not the functioning but the morphogenesis of the organism to see clearly the vanity of these efforts. Let's return to the example of the Venus flytrap. Like all multicellular organisms, it derives from a single cell. Its trap-structures, just as much as the motors of these structures and all of its other structures, derive from this sole initial cell, which developed by division and then by a complex play of interactions between evocators, inductors and competent tissues. Can it be said that the action of a biological inductor on a tissue which will become the trap organ is also analogous to that of a mechanical stimulus on a fully formed trap? It is quite clear that the cells or tissues which, in the course of development, form trap organs are not themselves able to be induced to differentiate themselves in the same way that the adult organ is induced to function. The fully formed trap visible in space is, to use a metaphor, 'competent' to trap flies. But the competence of a tissue, which reacts to an inductor by differentiation and formation, is profoundly different. It is not, by definition, legible in a structure *for this competence bears precisely on the formation of a structure*. The expression 'morphological charge' is sometimes used to designate this competence; the Germans say that the reacting tissue is '*gestaltladen*'. In any case, this 'charge' or potential is, it must be admitted, of a profoundly different nature to that of the fully assembled trap. The future organs are not 'there' in the way that the adult Venus flytrap, whether open or closed around its prey, is there in space.

E. Let's now pass to the true signal stimuli, and to the psycho-biological order. When a subscriber[11] dials my number into a telephone, this composition, once completed, functions automatically like a triggering key and makes the phone in my apartment ring. But when I perceive the sound, it invokes a whole behavioural complex in me: I know what I have to do in response to a phone call.

The potential evoked by the signal is mnemic: the stimulus was a true, perceived signal, effective 'information', and I was competent, according to a knowledge [*savoir*] and a precise code, to respond to this information with a complex behaviour which, once evoked, unfolded by itself until the next confirming signal.

The vibration of a web when the prey comes into contact with it equally evokes, in the spider, a whole complex behaviour until the next confirmation. In the same way, finally, when the epidermis is put in contact with the annular cartilage, a tympan is formed. In all of these cases, there is an obvious structural disproportion between the 'cause' and the effect, and unlike case C above, this disproportion is not immediately legible in the trap-like structure of the receiver.

Classical attempts to reduce signal stimuli to pure cases naturally come down to supposing, on the one hand, a hidden structure – an invisible trap – in the receiver, and on the other, the stimulus-trigger acting like a key. The competence to respond to the telephone, like that of capturing a fly, would simply be a structure formed in the course of the anterior life of the individual or the species in the nervous system. But for both acquired behaviour and instinct, it becomes clear that this interpretation runs aground as soon as the facts are examined more closely. Despite their premature predictions of victory, the manufacturers of automata have not managed to imitate thematic perception with mechanical models, that is, compliant to stimulus in the general sense of the term rather than to a particular and concrete structure.[12] But given that it runs aground in every respect and without possible discussion for morphogenesis – since, once again, what is stimulated in morphogenesis is not a functioning but a formation – it is logical to interpret all signal stimuli as beginning with a non-spatial potential. What good does it do to represent the nervous system of a spider, for example, or a female stickleback, as a sort of electronic automata assembled in such a way that it reacts to a key – the zig-zag dance of a partner or a vibration in the web? This nervous system has in fact been *constructed* in the course of morphogenesis, and the action of chemical signals, inductors, cannot be assimilated to that of a key when what is at issue is precisely the explanation of the formation of the keyhole.

A signal must therefore be considered, if not as always perceived in the strict sense – an embryonic tissue not having sensory organs – at least as awakening what must really be called a mnemic consciousness in the respondent, whether this is human, animal or an organic tissue. A signal is information in the psychological rather than the cybernetic sense of the word: it is necessarily 'perceived', in the broad sense of the word, in its signifying or expressive form, and this is why, unlike a key, it can act in its absence or lack just as well as its presence, for example, in organic need. This is why it is spontaneously generalised, as the facts discovered by Pavlov demonstrated a long time ago. And this is also why it can be incomplete, simplified, sometimes even overnormalised[13] or imitated by a lure; stimulated consciousness 'semantically' or 'mnemically' reconstitutes the initiated complete theme.

F. In passing from signals to signs, we pass from psycho-biology to psychology in the strict sense, and even – or so we would have been inclined to think before Frisch's discoveries – to human psychology.[14] To communicate by signs is to count on the consciousness of the correspondent, on a competence of the same nature as the competence which responds to signals but which is no longer uniquely mnemic and which involves a great deal of comprehensive improvisation in response to the analogous improvisation of emission. By imitative gesture, it is possible to suggest to someone with whom a language is not shared that they should come closer, stop or lie down. This relies on the existence of reconstructive initiative on the part of the interlocutor, in whom a consciousness that tends towards the meaning of the message must also exist. A sign, or a message that uses signs, in contrast to a signal, not only conveys information – it has *its own informational content* which must be perceived, this time in the strict sense of the word. The specific structure of the content must be transmitted. The sign is not a simple psychic trigger. In a language like the dance of the bees, if the bee finds a source of food less than fifty metres from the hive, it will dance in a circle, indicating no direction. As J. B. S. Haldane notes, this dance is an *Auslöser*[15] in Lorenz's sense, that is, a signal stimulus and not a sign. When this distance is exceeded, the bee will make use of true signs, or, if you like, its 'message'* possesses informational content: the 'waggle dance' describes a direction, and the number of dances per minute, the remoteness of the food.

In reality, it is often very difficult to absolutely separate signs and signals, and a sign always tends towards being conventionalised as a signal, as linguists and the contemporary theorists of information and messaging are quite right to point out. This is because the transmission of a message through pure signs, which are both non-conventionalised and have not been transposed into discontinuous signals having a value of all or nothing, is much more likely to be subject to error, and these small errors of transmission accumulate. A discontinuous and conventionalised signal, by contrast, can be sent over and over again without being deformed. Mandelbrot[16] uses the example of a message transmitted by the admiral of a flotilla from one boat to another, describing an angle of direction. If each boat *imitates* the given angle, errors would threaten to accumulate. If on the contrary, the angle is first rounded down to, for instance, one of seven possible directions (0, 45, 90, 135, etc.), the receiver – who is presumably competent, that is, familiar with the convention – can *repeat* rather than *imitate* the message and correct for small errors (by repeating 45 and not 44 or 46, for instance, even if they perceived 44 or 46). Words or formulae completely composed of linguistic messages are discontinuous signals even if the message as a whole transmits original information; it is much easier, as we know, to repeat a commercial letter over the telephone without error than a sequence of letters or numbers or, above all, a continuous sequence of sounds.

From this point of view, the messages-signs of the bee are less conventionalised than human language. The transposition of distance into a number of 'rounds' per minute is continuous and approximative. The indication of a direction by the waggle dance in a straight line is less conventionalised again. On a horizontal plane and in sunlight, it describes the direction relative to the sun. On a vertical plane and in shadow, the bee transposes, taking the angle with the vertical as the angle towards the sun. But as E. Wolf suspected and as subsequent experiments with other insects have shown,[17] there is no convention to be found here: for insects, the sun or the vertical are the privileged directions and are immediately equivalent. We discover here the thematic and spontaneously generalised character of all signal stimuli. The 'language' of the informant bee is 'orienting behaviour' rather than 'narrative behaviour'. It makes excited movements, turned towards the future and not the past, movements that its companions must participate in to understand. This conduct is closely analogous to that of the 'ant-centres of excitation' described by D. W. Morley. It only serves as an accessory of information. It is a movement of ritualised intention and in this way becomes a medium of communication, as Tinbergen and Armstrong have discovered in so many examples throughout the whole domain of instinct. The language of bees is equally close to the 'mood language' [*langage d'humeur*][18] that Lorenz describes in relation to jackdaw flocks,[19] where the call 'Kia' signifies 'I am in the mood to be far from the nest', and 'Kiaw' signifies 'I am in the mood to return', to which the individuals of the flock are respectively stimulated contradictorily, and then – by a 'recruitment' analogous to the recruitment of neurons in cerebral activity – unanimously.

SIGNAL STIMULI AND AGENT STIMULI

Broadly speaking, and setting aside language and human techniques of communication, along with certain aspects of the language of bees – the use of signs by bees and human beings possessing, incidentally, the significant value of leading to the recognition of the direct heritage of the sign and the signal stimulus and in turn the psychic character of signal stimuli – we can say that the biological form of 'efficacity' par excellence, in sequences of formation and behaviour, is the signal stimulus.

In the domain of what Viaud calls 'true tropisms' – for example, the orientation or locomotion of an organism, or even a fragment of an organism, towards a light source, a cathode or merely upwards when an organism *automatically*, *non-adaptively* and to the *greatest extent possible* pursues its provoked behaviour – it is often difficult to interpret a stimulus as a signal. It rather appears to be an agent stimuli, that is to say, a true cause. At issue here, however, is a limit phenomenon, and it is no longer today a question of seeing in it a fundamental phenomenon and relating it to instinct. Already in the case of the 'pathies', or adaptive repulsions (for example, flight towards a light or an extreme temperature) – in which animal behaviour, far from obeying the law of maximal excitation, succeeds in subtracting it from the action of the stimulus or in carrying it towards an optimum or 'preferendum' of excitation – the stimulus is certainly a signal stimulus and not an agent stimulus and must be 'perceived' (in the broadest sense) by the organism. There is such a gradual evolution from 'phototrophism' to 'orientation by the light of the sun used as a guide' that doubts arise concerning their strictly causal character and even concerning 'true tropisms'. If it is true, as Viaud says, that 'light first guides animals phototrophically', and then that 'they guide *themselves* by it', it is tempting to believe that guidance *by* light, even automatically, is already just as much an action *of* light on the organism as it is a response of the organism simply considered at a more basic level. The only purely mechanical action of gravitation on an organism takes place if the organism is in free fall – but if it is oriented upwards by a tropism, then the beginning of the fall, by definition, only acts as a signal.

THE VARIETY OF SIGNAL STIMULI

The true principle for the classification of signal stimuli is found in their diverse relationships with formative 'vertical' melody. A signal can be a trigger and initiator of an entire sequence, or it can confirm and relaunch. Formative or instinctive melody can, moreover, be relaunched by 'proprioception' – the end of a first phase serving to signal the beginning of another phase (for example, the end of the *zig* in the stickleback's dance triggering the *zag*) – or by 'extroception', the first phase having brought about a specific partner or a specific stimulation. In general, a stimulus, whether a trigger or a relauncher, can be internal – as is the case in the general sensitisation by a hormone, responsible for what is known as spontaneous behaviour, or in reductions of external stimuli down to the threshold of efficacity – or external, generally through sensory organs and nervous transmission. It must be emphasised that the effects of reductions to a threshold of efficacity of an external stimulus are not mechanical. In instinct, in any case, they express perceptual 'valorisations' of the stimulus-object, which appear as more and more 'pressing'. An internal stimulus triggers an activity of exploration, or searching, or pursuit. External stimuli can themselves be triggers, directors or orienters. For example, fledglings in the nest, still blind, are stimulated – or triggered – to raise their beaks upwards by a gentle agitation of the nest, which thus 'valorises' the vertical direction, gravity functioning as an orienting stimulus. Once they are capable of seeing, the sight of the mother, or of an equivalent lure, first acts as a pure trigger and not as an orienter. For a brief period, the fledglings stretch out their necks when they see the lure but always do so upwards, whatever the position of the lure. From then, the visual lure acts as both a trigger and an orienter at the same time.[20] It is remarkable that in the course of morphogenesis, stimuli are also found – not only triggers but orienters. Thus when the leading edges of developing nervous fibres make their way towards embryonic muscles in order to innervate them, it is likely the muscular cells emit orienting substances and that the nervous fibres are guided by these emissions to their path through the tissues.[21] Forms which are not strictly specific – like the arrangement of small veins, capillaries or the secondary branches of nerves – must indeed be guided in their formation by orientating stimuli. But since there are possibilities for both regulation and regeneration in the course of the most fundamental formations, it is also necessary, even for those formations in which the specific melodic theme is predominant, for a certain guidance by stimuli according to the existing circumstances to be brought to bear and to adjust the elaboration of this theme. And in fact, J. Holtfreter's meticulous observations in 1943 and 1944 on cellular movement during gastrulation directly reveal this guidance. The superficial cells which are displaced towards the blastopore and then absorbed within its interior (in the amphibian egg) are guided by the non-cellular substance which covers the egg, just as migrating colonial amoebae are guided by the material trace which is left by the leading amoeba. Once they pass into the interior, they maintain contact with the covering substance, remaining connected by a long foot which pushes on this surface, giving it the shape of an elongated pear.[22]

It concerns, therefore, a guidance – and even a 'sought' and maintained one – rather than an impulsion.

TRIGGERING SIGNALS AND INDEXICAL SIGNALS

The effects of internal and external triggering, orientation and guidance are quite often closely intermingled precisely because triggering, here, is not mechanical but mnemic – in the biological sense of the word, that is, 'bearing on a potential'.

Consider, for instance, the migrations of salmon or eels.[23] Why does a fish, which has remained sedentary for months, abruptly set off in this way? External, meteorological factors are clearly in play. Eels leave for the Sargasso Sea above all in autumn, when the waters run high, a moment determined by the lunar cycle. But a physiological state is also necessary given that, for fish of the same species submitted to the same external influences, the direction of migration can be diametrically opposed. The physiological state is therefore distinguishing, not only of the migratory state but of the direction of this migration itself. In salmon, it is an overexcitation, and then a disequilibrium, of the function of the thyroid, which is first translated by a lustrous appearance and by motor agitation, which renders the fish sensitive to external influences. The long migration itself, at once triggered and determined with respect to its general direction, must be guided, above all in order to return to the natal river, by indexical-stimuli whose action is conjugated with that of general direction. And there can be no question, properly speaking, of a sort of composition of forces. Indices are 'means' for the animal; it must perceive and make use of them according to an instinctive goal. What are these indices? The question remains unanswered. The facts rule out the hypothesis of guidance according to a salinity gradient. The sense of smell and olfactory memory cannot provide a trajectory in the open sea. Eels, for their part, do not appear to have a sense of the geographical position of their distant goal since they cannot move in a straight line or, to be more precise, according to a segment of a large circle as a bird or an ocean liner can. Baltic eels maintain a constant angle in relation to meridians, and follow what is known as the loxodromic line: they therefore are much more likely to possess a sense of the local *direction* of their movement. But the physiological support and the indices used, in this sense, are unknown.

These indices certainly exist, however, instinct is not magic, and, like all organic life, it does not take place through natural means. It is not 'causalist' in the sense that signals are not impulsions. But index-signals are indispensable in 'informing', in the psychological sense of the word, the melodic unfolding of an instinctive act, which is for its part 'informed' in the etymological sense of the word, which is to say that it is not amorphous. It is this – entirely justified – sense of the disproportion between the index-signal and the complexity of the action that inclines the layperson, and even the intuitionist philosopher, to speak of instinct as magical or mysterious. What is 'magical', if we were to use such a word, is the existence of these potential themes without spatial support and the appropriateness of specific themes, despite the dispersion and diversity of their supports, modification with respect to their general nature – for example, the instinct of the male and the female, the instinct of the adult and the larvae to feed, of the neutered insect and the fecund queen, of predator and prey, insect and entomophilous flower. But actual modification, as a particular event, is never, itself, magical. It is not 'extra-sensory' in Rhine's sense, nor is it due to a 'sympathetic' intuition in Bergson's sense.[24] Actual modification according to signals is rather the 'non-magical' part of instinct, just as the accommodation of diverse partial developments effected by chemical factors is the 'non-magical' part of formation. And it is for this reason that scientific embryology, like the biology of instinct, has a predilection for the study of the factors of modification.

THE SIMPLIFICATION OF SIGNALS

One of the most distinctive features of signal stimuli is their tendency towards simplification. A signal is always as schematic as it can be when a clear expectation on the part of the signal's receiver renders an incorrect interpretation unlikely. The two taps in the bathroom have no need to bear all of the letters 'Hot Water' and 'Cold Water'. The signal is quickly simplified to H and C, or, for international clientele at a hotel, it is imprinted in red and blue. The same holds for red and green traffic lights. The same holds for the familiar means of social signalling: bell, siren, drum, etc. The same holds for the distinctive marks that signal party or nationality. In his experiments on thought, Ach emphasises the fact that a subject's expectations spontaneously produces this abstraction from stimulus.[25] If, for example, the subject expects the letter S as a stimulus, only the letter S is perceived when the stimulus is presented and the other letters are ignored. If a colour is expected, its form is not noticed, and vice versa.

The signal-message obeys in this regard a law different from that which governs the sign-message, which possesses its own informational content. As A. Moles[26] has noted, a sign-message, with informational content, must possess a certain redundancy in order to be understood: a lecturer or public speaker systematically diminishes the density of information of their topic for an unfamiliar public. Conversely, in all languages, whenever the competence of the general public rules out any chance of equivocation, expressions are simplified: we no longer refer to an automobile vehicle, but to a car or a ride.[27] Signal stimuli also follow the rule of extreme simplification in instinct. A stickleback or male red-breasted robin in a combative mood reacts, like Ach's subjects, to simplified signals, for instance, the lure of a red belly signifying 'male rival'. Hungry fledglings react to a stick held above the nest.

It is natural to put morphogenetic stimuli in the same category as the ensemble of 'chemical messengers' in the same category: they are, it would seem, signs simplified into signals, which are themselves already simplified. In order to induce a competent tissue to form a neural canal, a banal introduction of substance, which does not need to be supplied by a living inductor, will suffice. Outside of the normal active substance – the key itself – there is always, as Dodds has insisted, a large number of simplified alternatives and even master keys. The extraordinary universality of hormones, like prolactine and auxin, has struck every biologist. Prolactine is a hormone related to lactation and maternal instinct in practically all mammals, but it is also prolactine that provokes the instinct to brood in pigeons and the secretion of 'pigeon milk' produced in the crop.[28] Auxin, an extremely banal substance which can be produced through the synthesis of analogues is, in plant life, a sort of 'all-purpose stimulus'. It stimulates the development of the stalk, arrests the development of the roots, serves as a signal to orient the stalk upwards and the roots downwards, assures the dominance of the principal stalk and makes no longer active leaves fall. In short, its morphogenetic role is enormous.

> It is difficult not to think of an analogous phenomenon in semantics which linguists call 'restriction' but which leads to giving certain words a universal use precisely because each group of speakers take it in a specific sense according to their own preoccupations and degree of competence. Bréal has noted the varied sense of the word 'operation', used equally by surgeons, financiers, mathematicians, theologians, etc.[29] Vendryès invokes the word 'season', used just as frequently by the director of a concert hall as it is by a wine-maker, a tailor or a fisherman and even, it could be said, by anyone at all involved in retail or industry. The organic or psychic signals which trigger 'the neurulation operation' or the 'operation of formation of a tympan or lens' or, in an adult individual, the 'season of love', 'combat operation' or 'operation of rearing children', are kinds of all-purpose, 'universally specialisable' 'words'. It is striking that many of the names for hormones have been falsified in their etymology because the immense variety of their roles had not yet been discovered: prolactine acts on organisms which do not produce milk, and auxin does not always invoke 'growth'.

THE EVOLUTIONARY ORIGIN OF SIMPLIFICATIONS

A justification for the thesis according to which instinctive biological stimuli are fundamentally psychic and not mechanical in their mode of action can be derived from the law of the simplification of signals-stimuli. In effect, if the 'psychic' interpretation of signals-stimuli, which brings together signs and the facts derived from their simplification, is rejected, we find ourselves in a strange situation when it comes to understanding the origin of instinctive stimuli. Where do they come from? And how can their evolutionary origin be understood? Take for example the instinct of numerous birds to cower while being circled by birds of prey. By parading various cardboard simulacra in front of farmyard birds, Tinbergen showed that the effective releaser* was 'the silhouette of a bird with a short neck'. Here is Julian Huxley's commentary: 'The instinct to crouch down appeared, at first glance, to require a Lamarckian explanation' – that is to say, a 'psychic' explanation of the kind that we are advancing – 'until the day when Tinbergen showed that a cardboard model could trigger the reaction'. The neo-Darwinist J. Huxley therefore concludes that fortuitous mutations have inclined the nervous system of gallinaceous birds[30] of this sort such that a figure, itself without any psychobiological signification, triggers the reaction of huddling down and that these mutations have been conserved by natural selection because the figure in question is found, by miraculous luck, to resemble that of a bird of prey in flight. In the same spirit, the behaviour of a spider when it rushes towards the vibrations that its prey makes in its web could be 'scientifically explained' by showing that a tuning fork is an effective stimulus. Such interpretations are inadmissible. The simplified lure clearly does not present the 'primitive mechanical kernel' of the effectiveness of the releaser*. Or if so, why couldn't we also claim that, in the human species, what is known as a 'skirt-chaser' is a male endowed with a tropism or a nervous reaction towards skirts and that the discovery of this tropism constitutes a scientific explanation, purified of all recourse to psychism, of sexual instinct?

What Tinbergen's studies demonstrate, to the contrary, is the signifying character of the evolution of instincts and their releasers*. Instincts are constituted *through veritable semantic displacements*, and they have evolved in the manner of a language. Tinbergen and Lorenz were capable of developing a veritable *etymology* of certain animal displays. Their work, despite the residue of mechanist postulates, presents a decisive argument in favour of the thematic character of all biological development.

It is often possible to come upon the progressive restriction of the sign to a signal at work in instinct. Just as E. S. Russell has noted, it is completely artificial to start with stereotypical reactions to stimuli in order to then recompose the behaviour of an animal, as naturalists often do in the laboratory. In reality, the animal always tries to do something specific: to construct a nest, to feed its young, to return to its normal habitat. This action, as a signifying whole, is often decomposed into stereotypical responses to precise signals even though the animal ordinarily remains capable of regulation and of subordinating variable means to a unique end. If, for instance, a young eel is isolated in a fish tank, it does not manifest its striking migratory possibilities but only a banal rheotropism, and for good reason. Must migration be explained on the basis of a composition of derivative, automatic rheotropic responses? Or should we not instead begin with migration in order to understand that in a tank, the fish tries its best to act in accordance with the general 'migration' theme? If the method of composition is adopted, impasses are very quickly run into. According to its age or physiological state, the eel will prefer to swim either with or against the current. In the same way, a bird takes a worm into its beak and swallows it in a sequence of 'sequential acts'. And yet, if it is in the process of finding food for its young, it keeps the worm in its beak despite the fact that according to the doctrine of sequential reflexes, swallowing would automatically follow the stimulus that is constituted by the presence of the worm in the beak.[31] In all of these cases, the general theme of the action that is underway dominates and controls the particular acts and responses just as the general morphological theme dominates and controls the varied effects of auxin on the stalk and the roots. The 'code' of signals was not learnt in any mythical school, but it is living in the theme in the course of its fragmentation, and it is developed on the basis of this theme.

THE ELEMENT OF CAUSALITY IN SIGNAL STIMULI

Signal stimuli are biological means. Now, every means clearly possesses an element of mechanical efficacy without which it would not be a means; there is always a reason according to which one means is used rather than another. There is no doubt, for example, that one part of the efficacy of hormones is purely chemical. The heat and light of spring act both as causes and as signals on plants and animals. A signal as such is arbitrary, but it could not be employed as a signal if it did not possess an immediate causal efficacy. The sound of a siren which is a signal to workers gives a passer-by a fright; a signal would not have been chosen if it did not have immediate effects of this kind.

We rediscover here a universal feature of living forms, just as with forms created by human technique: they are modelled *as closely as possible* on the physical laws that they nonetheless dominate and make use of. All organisms, for example, have had to adapt to the physico-chemical laws which make their nourishment possible. The important fact is that this adaptation is always the acquisition of a competence to respond in a specific manner to the action of the milieu which, therefore, is not a mechanical cause but the signal stimulus of a response.

Chapter 6

Competence

The absolutely decisive character of 'competence' is easily demonstrated. The term passed into the current usage of embryologists with Waddington. It designates the state of reactivity on the part of an embryonic tissue which allows it to respond to what is known as a morphogenetic stimulus with a differentiation in a determined direction. Through an inevitable extension, it is used to designate, beyond development, the state of reactivity in any organism or organic tissue to any stimulus, a hormone for example. Now, strictly speaking, the state of 'competence' is not observable. It is not enough, for instance, to look at the external tissue of a gastrula in order to see that it will react by forming a neural canal as soon as it comes into contact with a primary organising stimulus. Its behaviour can only be predicted by analogy and can only be guaranteed to have taken place when there is differentiation, which can itself be observed. In turn, this is why the metaphor of competence is required – someone's competence is only indirectly observable on the basis of their acts.

THE PSYCHIC CHARACTER OF COMPETENCE

We might venture to say that whenever embryologists speak about anything to do with competence, they display a sort of bad faith, and a clumsy bad faith at that. The word is taken, in a remarkably unscientific fashion, as synonymous with the words 'power' or 'potentiality', which simply indicate a possibility and not an effective power. But what is competence if not an effective power? And what is an effective power if it is not interpreted according to psychological experience? As we have seen, to speak of the 'competence' of a trap which functions if it is triggered, or the 'competence' of a domino to fall if it is knocked over, would be an abuse of language. In a machine, it is easy to distinguish the motor from the assemblage of parts required for a certain functioning. Once a machine is coupled to its motor, stimulus can set it in motion as many times as you like; it is enough to shake a watch a little in order to set it in motion once again.[1]

In a living tissue, there is no possible separation between a motor and an already present structural assembly, competence bearing precisely on the formation of a structure. Competence is nothing if it is not analogous to a psychological assembly, to a latent knowledge [*savoir*] and not to a structural assembly. It is analogous, in the final analysis, to the term 'task'. When the alarm goes off, the worker gets up and gets dressed; with blurry eyes, he regards the clothes in front of him, brusquely prioritised in terms of 'getting dressed the fastest'; new relations are established and the man in pyjamas transforms himself into a dressed man, and then into a man taking the bus. In Watt's experiments in psychology, the psychological presence of the task, whether conscious or subconscious, transforms a perception into a self-fulfilling signal.[2] The same is the case in the biological execution of a competence: the stimulus is valorised as a signal; a situation is valorised by the 'task'. New relations are created according to evoked knowledge [*savoir*], and structural modifications become observable.

This is no gratuitous theory arbitrarily opposed to mechanist theory since, in the case of instinct, we come upon the psychological character of competence in a concrete form. As E. S. Russell puts it, stimulus is perceived as valency, and in turn, every situation appears to the animal as a system of dynamic valencies, as relations to be established or broken. The spider perceives the vibration of its web as a call [*appel*], the centre of the vibration as the prey towards which to go, then as prey to be paralysed, etc. In the same way, the execution of a morphogenetic competence is necessarily a 'valencing', an active assembly of new relations according to 'the task at hand' but which has not yet been completed. Either competence is not morphogenetic or it is psychological in nature. There is no middle way through this dilemma.

Embryologists voluntarily delude themselves when they imagine that the vocabulary of 'stimulus' and 'competence' allows them to fundamentally condemn the metaphysical and anti-mechanist vocabulary of Driesch, his 'entelechy' and 'potentialities'. The vocabulary of 'stimulus' and 'competence' is indeed preferable to Driesch's not because it allows us to return to 'more scientific' (read 'mechanist') conceptions but because it clearly and necessarily implies a reference to psychism. It is indeed better to make use of a psychic interpretation by analogy with experiments in psychology than to a mysterious and metaphysical principle. But to entertain this illusion is to hope that 'competence' can be interpreted in terms of chemical phenomena, in the same way that it was possible to isolate the element of stimulus due to chemical efficacy. Needham and Waddington consider the incapacity to define competence to be a provisional lacuna and hope that a day will come on which it will be related to the actual presence of specific proteins in the tissues under examination. But this is to purely and simply return to preformationism, to deny morphogenesis and to make of visible form the amplification and mechanical effect of the 'smallest chemical forms' in the nucleus and the cytoplasm. Needham recognises, with a feigned impartiality, that 'the state of the reacting tissue is equally as important as the nature of the stimulus provided'.[3] This 'just' allocation of fifty percent raises a smile; it is, moreover, difficult to see how this mathematical proportion can be reconciled with another, advanced by the same author: 'To the extent that the chemical specificity of stimuli decreases, the greater the weight that must be given to reactivity, that is, to the competence of tissues'.[4] The deception of biologists each time that they discover the extreme banality of a stimulus in which they had hoped to find the key to a formation is nevertheless difficult to hide. Bonner is more honest: he compares the organism to a garden in an arid country, and the stimulus – specifically, auxin – to the water from a garden hose.[5] By directing its allocation, the gardener is able to make a certain plant grow while leaving another dormant; by overwatering, he can even arrest the development of the plants that require very little water. But what he cannot convince himself of is that the nature of the water is 'just as important' as the specific properties of the plants' tissues.

THE NON-SPECIFICITY OF INDUCTORS

The enormous importance of reactive tissue, and, consequently, what effectively remains as the complete and inviolable character of the morphogenetic mystery, can be established by innumerable experimental arguments. Principal among these is the non-specificity of inductors or stimulators. When Spemann baptised the privileged region which induces the neural canal and inaugurates differentiation as a whole – and which, when implanted on the side of an embryo, will produce a secondary embryo – with the name 'organiser', he believed himself to have isolated the vital key. When Holtfreter discovered that the same material produced the same effect even after being killed and boiled, and then that adult tissue taken from a whole range of animals acts in the same way, and then that a whole series of chemical substances were equally effective – themselves also found in feminising hormones and carcinogens – it became difficult to maintain the ambition of a word like 'organiser'. Sex can be made to change orientation – for instance, that of a green spoonworm: the masculine sex is normally produced when an asexual larva comes into contact with the proboscis of a female as the result of a wide range of banal influences: traces of copper or glycerol, a certain concentration of CO_2, etc. Even in normal development, a sort of indifference to stimuli is often manifested – provided that there is one. The plumage of birds is affected, depending on the situation, by male or female hormones, by hormones from the anterior hypothalamus or by hormones from the thyroid.[6] From one species to another, the 'inductor-induced' order is reversed, proof that what counts is the form finally attained much more than it is the order or nature of the 'uncoupling'. Thus, even though the optical vesicle in vertebrates is an outgrowth of the primitive brain, which induces the formation of the lens on the ectoderm, in cephalopods – whose eyes nevertheless greatly resemble those of vertebrates – the optical cavity develops, to the contrary, on the basis of an invagination of the ectoderm, and the optical ganglion is a differentiation dependent upon a stimulus coming from this invagination of ectodermic origins.[7]

As de Beer has shown in a general terms, *the structures that result from the same inductors are not necessarily homologous, and homologous structures can result from different inductors.*[8] They are not necessarily born from the same situations in the embryo or the same primary tissues. The specificity of forms and behaviours displays a certain obstinacy; they are realised not *in accordance with* but *in spite of* the variety of stimuli in play. And the same stimulus can be applied to different competences, their forms being different.

> Spemann and Rotman have interchanged the material of the lens between two species of tritons: the dimensions of the lens that develops always conforms to what it is for the donor of the graft and not the recipient, which contributes the inductive stimulus. This obstinacy in the maintenance of specificity is all the more remarkable in light of the contrast with the docility of the tissue in producing a different organ if it is placed at a sufficiently early moment in a situation in which it will be subordinated to different inductions. In the same way, while it is possible to interchange the tissue which would normally form the mouth between two species of amphibians, it will always respond in the same way and, as Spemann puts it, 'in its own language'. The inductor of the mouth in *Anura* normally forms a horned jaw, but if *Urodela* tissue is introduced into the buccal tissue, the inductor of the host will induce teeth and not horns, and vice versa (Schotte's experiments).[9] It is as if, Spemann says, the indication given by the 'buccal structure' stimulus was completely general and that the tissue responds to this indication according to its *specific* competence. This leaves little doubt that the indicative stimulus which sends the signal 'form the buccal structure' to the grafted tissue is only a banal chemical substance.

To speak of a 'language' of signals is just to speak metaphorically. But this is even more reason why the same cannot be said for 'response', which is indeed, if not a language, then at least a complex expression of a competence which is not materially inscribed in the tissues but psycho-biological in character. It is remarkable that this specific competence manifests itself equally well in psychological rhythms and instinctive behaviour as it does for formations in the strict sense of the word.

> If, for example, the material that forms the heart is grafted between axolotls of two species whose hearts beat in different rhythms, the rhythm of the grafted heart will conform to the rhythm of the 'donor' (Copenhaver). What is more, Giersberg's truly remarkable experiment,

which exchanges by grafting the embryonic brains of two species of amphibians (*Pelobates fuscus* and *Rana arvalis*)[10] at a primitive stage, the adult animal that thus possesses the brain of the *Pelobates* in the body of a *Rana* displays the tunnelling instinct characteristic of the *Pelobates*.[11] By contrast, if the intergrafting takes place between two animals of the same species, and even if the experimenter exchanges organs which are in different developmental stages, the graft will in general adapt to the 'time' of the host. Twitty and Elliot, for instance, remove both eyes of a medium-size *Ambystoma*[12] and graft them onto smaller and much larger individuals respectively. In both cases, the graft adapts its developmental rhythms, catching up or lagging behind until it harmonises its size with that of the host.

Closely related facts appear in metamorphosis. It is of course specific, possessing a quite variable character and amplitude across species, even though it is governed by the same hormone in all amphibians: thyroxine, whose essential element, iodine, is sometimes effective on its own. Metamorphosis is of additional interest in that it clearly demonstrates that for the animal undergoing metamorphosis, different tissues react to the same hormonal signal according to their respective competencies. The moment of metamorphosis for an individual can be brought forward or delayed by the introduction of thyroid extract or the privation of iodine. The moment of metamorphosis of a given tissue can even be changed by grafting the tissue into a younger or older animal. What cannot be changed is the nature of the reaction of the animal or the tissue. Each responds in its own way: limb buds grow, the tail degenerates, the gills atrophy, the skin becomes thicker, the intestine shortens, etc. Champy has demonstrated that at the moment of metamorphosis, the cells of two *contiguous* tissues – for example, the epidermis of a bud for an anterior limb and the coating of a gill slit – behave in contrary ways, the one actively multiplying, the other degenerating. The boundary is abrupt. In this case, competence thus appears still more predominant than the competence of plant tissues in response to auxin. There are no proportional effects: though a hormone certainly has a threshold of concentration, the response is not proportional to the stimulus.

The fervent faith of biologists in physico-chemical explanations when confronted with such facts is reassured by the thought that, after all, the same physical cause can also act on diverse physical materials and produce diverse effects: heat makes butter melt and egg whites harden even though they are next to each other in the same pan, just as the cells observed by Champy are next to each other. Even given all this, it should not be concluded that heat is a signal and not a cause such that butter and albumin are psychologically competent to give a specific response. The situation of diverse organs in metamorphosis, or in development more generally, is completely different. If thyroxine affects them as heat does butter, by what miracle are all the diverse effects so well adjusted that the modifications of each tissue converge with the modifications of each other in order to produce a harmonious organism?

What justifies the illusion of experimental scientists is the fact, of course, that they most often observe the *beginning* of differentiation. In both ordinary development and metamorphosis, these beginnings are very often very simple, structurally speaking. The primary inductor begins by only producing the start of a gutter, a simple fold; at first, thyroxine simply accelerates cellular division in certain regions. There is nothing astonishing in admitting that simple chemical effects are sufficient here. But since this is only a beginning – the gutter will become a whole nervous system, and the tadpole a frog – it must be admitted that the 'cause' was a signal and that the 'effect' was a complete response. The contrary action of a hormone on two neighbouring tissues must rather be compared to the contrary action of the same siren on a group of outbound workers, for whom it signals 'end of shift', and on a group of inbound workers, for whom it signals 'start of shift'.

For the tissues of the gills and the tail of an amphibian in metamorphosis, the hormone is even the signal 'end of existence'. What is most remarkable is that, in this 'competence to atrophy', the tissues take charge of their own elimination. For a long time it was thought that the atrophying of the tail during metamorphosis was a simple physical effect caused when the growing vertebral column blocked certain blood vessels. But subsequent experiments have shown that it is rather a question of a 'competence to atrophy' – if the tail is grafted onto another part of the body before metamorphosis, it atrophies when the time comes. Conversely, if limb bud tissue or even the primordium of an eye is grafted onto the tail of a frog tadpole, these tissues subsist even though the tail itself disappears – and, in Schwind's strange experiment, the grafted eye is displaced to the extent that the tissues supporting it are reabsorbed in order for the eye to be finally located on the frog's sacrum, where it resembles an aft lantern.

COMPETENCE AND APPRENTICESHIP

What distinguishes biological competence from human competence is that it is not acquired progressively and through the efforts of an apprenticeship. It appears and disappears; it is as if it is given to tissues and then withdrawn from them. Before gastrulation, the cells that will become the ectoderm are not yet competent to form a neural canal; on the other hand, if an ectoderm that formed a nervous system is isolated, it loses its competence at a certain moment. Before a certain age, thyroid hormones have no effect on the tissues of a tadpole. But later, the tissues 'know' how to respond to it without having learnt anything. Certain amphibians, like the axolotl, maintain the capacity for metamorphosis but partially lose the necessary stimulus while others (*Necturus*)[13] keep the stimulus but lose the competence. But if biological competence does not resemble human competence, it at least resembles a non-mechanist psychological competence – a competence for an instinctive behaviour, which is incontestably psychological, is likewise given in animals without a prior apprenticeship and can be lost if it is not aroused from time to time. Competence is a specific memory which can remain dormant, not a piece of machinery that does not function without a motor.

AN ATTEMPT AT AN ANALYSIS OF COMPETENCE

Biologists have nevertheless tried to develop an analysis of competence as if they were developing an analysis of stimuli – without any notable success. They have only discovered phases in the response, each of which, incidentally, possesses the character of meaningful behaviour, only to differing levels of integration. These experiments even have the great virtue of emphasising the analogy between a competent tissue which responds to a hormone and a colony of bacteria (or amoebae) which responds – or respond – to a chemical signal. What will actually happen, Umanski and Holtfreter asked themselves, if a competent tissue (for example, the ectoderm at the moment when it is competent to form a neural canal in response to an inductor) is cut up into cells the size of a hair, producing an amorphous cellular mass? Put into contact with an inductor, this mass tries to respond: cells differentiate into nervous cells and even sometimes manage to form a more or less typical structure, grouping together nervous canals in what looks like a normal fashion. Nevertheless, and particularly if the inductor has not been taken from living tissue, the histological response of the competent tissue, reduced to the state of an amorphous cellular crowd, succeeds better than its morphological response. This seems to indicate that the stimulus inductor acts first on cellular structuration, which in turn – when the cells have not been artificially put in a state in which it cannot do so – produces the morphological structure of the whole. Likewise, in a colony of amoebae, the structure of the whole is, to start with, less a competence of the whole than the result of a competence of certain individual amoebae induced by signal stimuli to group together and move around in a certain fashion.

The link between properly cellular competence and general morphological competence appears very clearly in many other domains. Certain species of salamanders have a uniform pigmentation spread over all of their bodies, while others have a line of pigmentation concentrated along their flanks. Twitty has shown that this morphological character depends on an instinctive competence of pigment-bearing cells.[14] If they are cultivated in vitro, they show a migratory instinct to either disperse or reassemble, according to their species, like the amoebae of *Dictyostelium*. Twitty has even shown that just like amoebic colonies, they emit a signal substance, to which they respond, in accordance with their competence or instinct, by either dispersing or grouping together. The overall pigmentary design of the animal, therefore, results from the individual competence of the pigmentary cells. In the development of the nervous system, a general competence of the tissue must certainly coordinate the cellular competencies given that the modifications there are much more complicated than reassembly or dispersion. But the experiments of Umanski and Holtfreter must be understood to show a distinction between cellular competence and morphological competence in the strict sense. This does not – to the contrary – rule out the conclusion that all competencies are of the same nature, instinctive and psycho-mnemic.

Chapter 7

Autonomous Procedures and Regulated Behaviour

The nature of formative competence can be clarified by a thorough study of instinct. That behaviour is not a supplement to formation but its very principle can never be emphasised too strongly. It is clearly the nesting instinct, in all its variations, which directs the formation of the nest, the instinct to weave that directs the formation of the spider's web. It is the instinct's own rhythm that directs the particular forms of its products.

ACTIVITY AND FORM

Spatial form is always second in relation to an activity. The bird makes its nest, and embryonic cells of a certain phase make the forms of the subsequent phase. A mollusc makes the spiral of its shell like a kind of nest whose regular development is inscribed in space. The designs of a mollusc shell are the work of cells at the edge of the mantle, which secrete the pigments which are incorporated into the shell. If certain shells are more active than others, this difference in activity will be translated into bands of more intense colour spiralling along the contour of the shell. If certain cells work according to a periodic rhythm, this rhythm will be translated into transversal bands around the shell.[1] In every domain, the spatial aspect of form is the product of an activity, and of the 'form' of its rhythm or its melody. Unlike transversal bands on a shell, longitudinal bands immediately depend, it is true, on the spatial disposition of the cells that produce pigments. But this disposition itself came about through active deployment.

This predominance of activity over spatial form is beginning to be universally recognised, from atomic physicists to sociologists, such that it can be said to be one of the distinctive characteristics of twentieth-century science. But in all of these same domains, scientists tend to assume as a result that spatial form is the pure effect of activity, an effect that is in itself arbitrary, meaningless and inconsequential. The excessive causalism of the nineteenth century continues to be combined with the most modern conceptions of activity. The fact that modern science no longer believes in

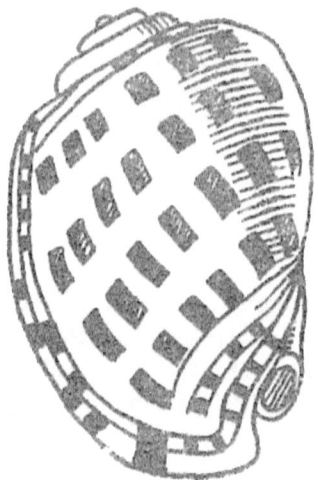

Fig. 7.1

forms but only dynamic processes sometimes leads to the attribution of a 'non-Platonic' character. This interpretation is problematic; it would be just as correct to speak of an intensification of Platonism. First of all, formative activities only result in a spatial form on the condition that they are not themselves 'amorphous'. Form, in a more subtle sense, is rediscovered in activity as the theme of an active structuration. But above all, the form resulting from organic activity is not arbitrary. Even superficial decorative patterns* quite often possess a meaning and an aesthetic or utilitarian value, functioning as camouflage, for sexual display or as displays of aggression. Their 'fortuitous fringe', like every fortuitous fringe of instinctive activity, also indirectly yields an aesthetic result. The notion of formative activity, so long as it is not confused with functioning in the fashion of the mechanists of the last century, implies that the future spatial form supervenes in activity as a directing theme if not as an already constituted model or precise plan for fabrication. The fact that the nesting instinct is, for those hoping to explain it, more fundamental than the spatial form of the constructed nest does not lead to the conclusion that the nest is a pure effect. It is clearly, to the contrary, the omnipresent sense of nesting activity.

We are freely inspired here by A. Portmann's remark, itself inspired by von Uexküll.[2] If a buried shard of glass were found bearing this design (figure 7.2), it would certainly be interesting to analyse the modes and materials of formation, the pigment of which it is composed and the likely instrument by which it was made. But the key would be to realise that the white spaces

Fig. 7.2

should be examined and that the word ZEUS is spelt out by them. To study the means and even the mode of execution does not exhaust the analysis of either the form or the formative activity, whose significance is greater than that of the matter involved. In a nest, what matters – as it does in a vase, as Lao Tzu says, or, let's add, in a host of organs acting as a receptacle or duct – is the cavity, the interior void, what is 'left blank', since it is this void that will shelter the eggs and hatchlings. The bird's activity, in building the floor and walls of the nest, comes *before* the spatial form is materialised; but the form itself, or rather its immaterial sense, is '*before*' active fabrication, being its theme.

The careful study of instinctive behaviour is therefore likely to reveal, to a much greater degree than the chemical analysis of stimuli or the materials present in a tissue, what morphogenesis itself is. If the fact that formative activity comes before realised form and, correlatively, the fact that this activity is truly thematic are taken as given, it remains to attempt to understand, above all on the basis of what we know of instinct, what constitutes the mode of this activity.

THE THREE MODES OF THEMATIC ACTIVITY

Any activity whatsoever can be thematic in three ways: (a) a builder constructs a house on the basis of an architect's plan; (b) a man 'mechanically' carries out a series of habitual gestures in the dark; (c) a shivering man draws close to a fire in order to warm himself. Let's also give some correlative examples: a pianist (a) sight-reads a piece of sheet music or (b) plays a piece he knows by heart or (c) composes, fumbling, a melody that expresses a vague feeling that he is experiencing. These three types of activity are equally thematic, but their modality differs. They can be characterised as (a) action according to a plan, (b) action undertaken 'mechanically', (c) action of improvised regulation. The first type of action is very rare in instinct. Only bees and perhaps ants are capable of being inspired by a theme of action indicated by the signs of their fellows. And it is scarcely conceivable in morphogenesis, at least without recourse to a mythical Platonism. The two other types are, by contrast, possible. Current theories of instinct are divided on precisely this subject. Broadly speaking, German and Dutch theorists (Tinbergen, Lorenz and their students) emphasise the role of 'autonomous procedures', while American theorists (Hull, Hebb, Richter and Lehrman) reduce the role of innate spontaneity to an extreme minimum and instead emphasise the role of improvised behaviour and learning by trial and error. The German theories are inspired in part by *Gestalttheorie* insofar as it believes in self-subsistent forms and not insofar as it reduces these forms to improvised equilibria. The American theories are inspired by behaviourism and cybernetics. Both approaches, incidentally, try their best to give mechanist interpretations to their discoveries, the result of an unexamined residual rather than under the constraint of the facts. If the discovered facts and their description alone are considered, and the mechanist interpretations are ruled out, the two schools both appear to be right in their respective conclusions. In instinct, the two types of action encounter each other and are closely intermingled; the same holds in morphogenesis.

AUTONOMOUS PROCEDURES

Let's first look at the category of 'autonomous procedure'. At the beginning of this century, two zoologists, Whitman and Heinroth, sought to theorise the relations between the different species of pigeons and the different species of geese and ducks with greater precision. They were looking for a morphological characteristic but found instead that the best classificatory guide was instead a type of movement, a 'motor pattern', common to all of the species of a genus and even an order, that functioned as an infallible criterion: 'It is possible, for example, to predict with complete certainty that all newly discovered species belonging to the order of pigeons (*Columbidae*) will drink in a certain way, making sucking movements to draw in the water, and that all species of *Anatidae* (geese and ducks) will draw oil from their preen gland by shaking their heads in a circular motion'.[3] This discovery was the origin of a completely new science, comparative ethology, which endeavours to compose a complete inventory of every type of behaviour belonging to a species and to study the embryology and development of the individual, their evolutionary origin, their displacements and homologies.

A motor behaviour of this kind possesses all of the features of a specific structure. It is innate, no more modifiable by its milieu than any Mendelian trait; it is even dissociable, through hybridisation, from the organic structures with which it is engaged: mutants of *Drosophilia* 'without wings' maintain the typical movements associated with drying their wings. This is not acquired through an individual apprenticeship, appearing in the individual by the convergence of 'determinations', like a complex organ. They often appear before the organ with which they are associated has finished developing. Thus a young gosling will hold its opponent in its beak in exactly the position where its wing would strike it if the wing were of adult proportions. It is even susceptible to 'original' development when a graft brings with it a competence for behaviour like structuration, just as Weiss's experiments on the development of grafted legs in *Amblystomae*.[4]

It is above all *autonomous*, which is to say that it has, in itself, a kind of procedural spring. Despite the fact that, normally, orienting signal stimuli are effective in guiding it, and that a signal stimulus is necessary to trigger it, sometimes the tension of the spring is such that the threshold of effectiveness is reduced to the minimum and the orienting stimuli are misrecognised. A fighting fish kept without a rival male will engage in attacking motions towards his female mate; the movements of combat will also be played out in a void, in an immutable order, each of its different components appearing in a more or less complete fashion according to an autonomous tension, and sometimes rhythmically repeated. Thus a captive starling catches sight of a non-existent insect and then engages in the movements of capture and swallowing. In the absence of any bird of prey, ducks go through the motions of flight and escape. In the same way, sympathetic pregnancy and the burial of food often take place in the absence of any object.

It is important to note that these kinds of autonomous movements and procedures are often absent in physiology in the strict sense, in morphogenesis, in, for example, the autonomous beating of the heart or the earthworm's crawling motion. Von Holst, for example, has completely isolated the nervous system of the earthworm, which he suspends in Ringer's solution.[5] This nervous system continues to produce the impulses that direct crawling with perfect coordination. In morphogenesis, the competence for a movement or a formation can also be manifested in the absence of normal stimulus. Thus in amphibians, the lens is evoked by a stimulus emanating from the optical cavity. If an experimental scientist prevents the optical cavity from coming into contact with the competent territory, the lens will not form, at least in the majority of species. But in the frog (*Rana esculenta*), it is formed 'by itself'. Correlatively, in the metamorphosis of the tadpole, the skin must be perforated to allow for the passage of the buds of the forelimbs. These morphogenetic facts concerning 'double assurance' have been contested, as have Lorenz's 'movements in a void'.[6] That they are questionable is easy to see given that it is impossible to absolutely demonstrate that no stimulus has acted. But whether the stimulus is infinitesimal or non-existent, it remains the case that quasi-autonomous procedures exist. In instinctive movements, a stimulus can obviously be hallucinatory: the starling goes through the motions of opening its beak to enlarge a crack and trap an insect even when no crack is within its reach; a dog observed by E. S. Russell 'buries' an object that was bothering it by pushing it under an imaginary earth with its muzzle.[7] In the same way, in the morphogenesis of mammals, the embryo produces the same annexes as the embryo of a bird or a reptile, notably an umbilical vesicle corresponding to the vitelline sac even though there is no vitelline to be put in the sac.[8]

'REGULATED' BEHAVIOUR

Let's now look at the category of 'regulated behaviour', improvised or acquired by practice. If the behaviours in the strict sense are considered in their full scope, rather than the isolated instinctive *movements*, for example, maternal behaviour or alimentary behaviour, the scene changes and the role of autonomous motor procedures, if it exists, appears in any case as secondary. A priori, an animal can be thought to have little chance of surviving in a complex and hostile environment if it can only engage in stereotypical movements even when they are more or less corrected by orienting stimuli. Of much greater importance are the directly improvised responses to external or internal stimulus, to the stimulus acting as a sort of irritant (drive stimulus*), not as a signal, progressively developed by conditioning and apprenticeship, or regulated by feedback* in relation to their sensible effects. The impression that maternal behaviour is correctly undertaken from the beginning without apprenticeship is erroneous. Thus there are the pigeons *which have already fed* which, if they have been injected with prolactine, set to feeding by regurgitating crop milk young birds who have been placed in artificial nests – young birds that, without prolactine, they would otherwise court or attack.[9] But if the birds have not already had a first experience, regardless of the injection of prolactine, they will neither approach nor feed the young. If they also do not attack, this is probably the result of an inhibiting effect of prolactine on the gonads. They seem to be under the influence of a great tension. But this uneasy tension does not know how to release itself through acts. In another experiment, even the experienced and prolactinated birds did not feed the young when their crop was anaesthetised. It therefore seems that the effect of prolactine is not to engender a sequence of stereotypical movements in the central nervous system but only to enlarge the crop, to produce a superficial irritation that provokes in turn its own acts of release. It also seems that the presence of the young is far from being an infallible signal stimulus and that the response is a matter of trial and error and must be learnt from neighbouring behaviours which appear to the animal as good 'regulators'.

It is obviously the case that a pigeon must have fed for a first time, or that a rat, even a first-time mother, will construct a nest in which to clean and suckle her newborn. But as Lehrman, a member of the American school, remarks, the fact that a pattern* of action is realised the first time an animal is in a biologically suitable situation does not necessarily prove that exercise has not played a role in its genesis.[10]

> Reiss and Birch's curious experiments eliminate any experience of carrying objects from birth for female rats and put collars around their necks while they are gestating, collars which prevent them from licking their own bodies and in particular their genital organs and which are only removed at the moment of birth. These rats do not know how to construct a nest and they eat their young rather than licking them clean. The transport of any objects whatsoever must therefore constitute the practice for the 'transport of objects for the construction of the nest' pattern, and the licking of the body must constitute the behaviour modelled for the 'cleaning of the young' pattern.
>
> Furthermore, it can be shown that the construction of a nest is related to the efforts of the rat to maintain the temperature of its body. So-called 'nesting' hormones do exist, but they only really act to increase the body's temperature to its minimum. In the case of the pigeon, it is even possible to reconstruct the sequence of laborious developments which lead to the first instance of parental behaviour. Diverse hormones, which inflate and deform the abdomen with blood, cause a local irritation which the smooth and cold contact of the eggs relieves: thus the bird learns to brood. A little after hatching, a tension in the crop produces a tendency towards regurgitation. The presence of newly hatched young accentuates this, and it is often observed that regurgitation happens for the first time by chance when the pigeon, cleaning the plumes of its chest, encounters the beak of one of its young. The feeding behaviour is thus organised. But the close proximity of the young is indispensable in the first instance; that the unexperienced pigeon who has not followed the procedure does not know how to feed the young placed in the artificial nest is easily understood.

Instinct, if there is such a thing, therefore appears as indissociable from a continuous apprenticeship, from a continuous and improvised effort to remove discomfort, and to maintain the sensations of relaxation. There are no automatic procedures or innate and stereotypical movements to be seen here.

Richter's investigations of the appetite of rats leads in the same direction. Take for example the rat's appetite for salt. The rat has no instinct whatsoever to absorb the salt that its organism needs; to the contrary, it functions like a very delicate chemical homeostat in order to maintain the optimal proportion of salt in its regime if, for example, it is offered salt solution in a concentration it can taste as an accompaniment to desalinated food. If the adrenal glands of a rat are removed, it rapidly dies due to an excessive loss of salt through urination. But if it is allowed to absorb salt voluntarily, it will keep

150 itself alive by *spontaneously* augmenting the ingested dose – even during the gestation period. Between saline solutions of various proportions, the rat chooses that which will best maintain the optimal proportion of salt, given its alimentary regime and its hormonal state.

MIXED BEHAVIOUR

The contrast between this type of behaviour, such as it is interpreted by the American school, and the first type, such as it is interpreted by the German school, appears abrupt. In the second type, central, innate, preformed behaviour, separable into acquired elements and independent of sensations and external stimulation – which, in Lorenz's conception, can only bring about a change at the level of details – cannot be found. And, properly understood, the American psychologists call the German interpretation into question even for behaviour of the first type. In their view, practice and apprenticeship are present everywhere. Doubtless, the innate exists, but it is inseparable from the acquired. There is no central pattern* of movement, entirely innate, always ready to be performed, present in the structure of the central nervous system, and triggered en bloc by a hormonal state or an external state. Peripheral sensations, or irritations and engorgements caused by hormones, provide more than orientations of detail. They are the motor of the act itself insofar as the animal tries to diminish tension or irritation. The animal is equivalent to a homeostat – to a self-regulating perfected automat with feedback*, not to an automat such as they were built in the eighteenth century or a barrel organ furnished with fully pre-written melodies. Regulation by organic 'proprioceptive' sensations plays a decisive role in instinct, above all when it involves an instinctive need. Respiration, for example, is an ensemble of movements coordinated by specialised nervous centres when it takes place normally. But as J. B. S. Haldane has emphasised, for an aerial, breathing animal submerged in water, or for a human being buried alive, the sensation of suffocation caused 151 by the abnormal accumulation of carbonic acid makes of the respiratory act an instinct which demands priority over every other act. It demands improvisatory activity: to return to the surface or to break through the soil or snow thus becomes as much the 'respiratory act' as normal inspiration or expiration. Sexual instinct is, correlatively, very often presented in psychological experiments as an improvised effort, a matter of trial and error, often aberrant, of releasing a tension and maintaining agreeable sensations that, like a sort of triggered motor sequence, is triggered once and for all.

Examining the matter more closely, there are many instances of 'proprioception' even in the cases cited by Lorenz and Tinbergen. The exact nature of the crawling motion of the earthworm or the swimming motion of the eel, for example, has not yet been entirely clarified. It is believed that contraction or torsion is propagated like a chain of reflexes, each muscular contraction depending on the proprioceptive stimulation caused by the contraction of the previous muscle. Then again, as we have seen, von Holst's experiments appear to refute this conception: a worm's nervous system continues to direct crawling even if it receives no proprioceptive sensation from the muscles or the skin. Locomotive movements in eels do not begin at the cephalic extremity before being propagated further back; a peripheral stimulation is applied, but this acts simultaneously on all of the muscles in its trunk. When it is weak, it forces the whole animal into a wave-like posture; stronger, and the rhythmic activities of all of its muscles begin simultaneously.[11] But other physiologists do not agree with von Holst, and it would seem that if all proprioception in a cat is supressed, or even in fish and amphibians (Gray and Lissmann), normal locomotion does not occur.

These difficulties and hesitations on the part of biologists are understandable given that, in the light of many human behaviours, 'Gestaltised' melodies and regulations by sensation of the obtained effect are intermixed in an inextricable fashion and are capable of substituting for each other. Writing, for example: in the child learning to write, writing is clearly regulated by sight. But a normal adult whose eyesight is too poor to read is nevertheless capable of writing correctly so long as he can roughly see the line and where to place the word. The motor melody which brings about the writing of a word is triggered in its entirety, like the swimming of an eel. The completely blind, however, can no longer write unless, like the wounded D'Annunzio, he makes use of thin strips of paper which allow for tactile guidance.[12]

AN INTERPRETATION OF LORENZ'S THEORY

What makes the conflict between the two interpretations so acute is that they rediscover, in a new form, an old, fundamental theoretical contrast: belief in heredity or the milieu, in preformation or in epigenesis. The American school believes in the complete interdependence between organism and milieu; it believes in the epigenesis of instinctive behaviour and reproaches the German school for being preformationist. We have deliberately kept the weakest part of the Lorenz-Tinbergen theory in the shadows: the interpretation of autonomous procedures by reference to specialised nervous mechanisms. These nervous mechanisms, centrally assembled, are progressively put into action, or made sensitive to the least stimulus, through the central accumulation of energy (nervous or hormonal) in the manner of the ancient forms of hydraulic automaton. It can even lead, under the effect of an abnormal pressure or a blockage of a normal outlet, to the activation of inappropriate mechanisms by this excessive energy. This is how displacement activities are explained, such as the smoothing of plumes or the digging of a nest in the middle of a fight.

Tinbergen believes he has found a confirmation of these views in the experiments of Hess and Brügger which, by introducing small electrodes into the diencephalon of living cats, can detect which regions an excitation stimulates, in perfect coordination with fighting behaviour or the search for food. He sees in this proof of the existence of mechanisms which receive excitations and redistribute them in the form of completely structured movements. Now, this interpretation – which exemplifies the old mechanical model of the trap or the barrel organ – has very little truth to it.

> The most recent experiments of Hess, A. Kert and Delgado (1953) have revealed that excitation of the same point in the diencephalon produces quite variable motor effects depending on the nature of the influx and that the same effects can be produced by the stimulation of different locations. More generally, we know that stimulation of motor centres at the level of the diencephalon or the cortex leads to such variable results that it would be difficult to imagine, in these centres, the sorts of sections that are always taken for stereotypical behaviours.[13]

Neurologists have admitted the difficulty of understanding the temporal order of a complex behaviour, like the pronunciation of a phrase by a human being or the execution of an instinctive act.[14] The old theory of an associative chain of reflexes is indefensible. In a verbal phrase, just as in a 'motor phrase', each word and each movement cannot be brought about by the sole excitation of a word or the preceding movement. This is clearly seen, for example, in the case of a lapsus and its compensation. A wife says angrily to her husband, 'I tyed a marrant!' instead of 'I married a tyrant!'[15] The main word 'tyrant' is preempted in the phrase so that 'marr', the part of the displaced word, must be lodged elsewhere as compensation. It is necessary, therefore, for a relatively timeless directive theme to preside over the deployment of the temporal sequence, beginning with the deployment of the syntactic schema. The 'monstrous' word 'ty-ed' is, grammatically speaking, a correct use of a past participle; that it is wrongly 'filled in' makes the 'verticalism' of the theme manifest, just as the Chinook language that interested Raymond Queneau does in which the words and their syntactic form are expressed separately. Vendryès has remarked that in spoken French – as in Chinook – phrases can be found in which the schema is given before their content: 'The robber, he never got 'im, did he, the copper?'[16] On this point, linguists are perfectly in accord with the neurologists and psychologists who have studied the various forms of aphasia[17] and who, according to Hughlings Jackson's 'verticalist' theories, have demonstrated that there exists a sort of cascading of determination in the formation of the phrase, from the abstract sketch [*ébauche*] to the 'morphemes' and 'semantemes' placed in temporal order.

Let's return to motor behaviour. Lashley has shown that for the pianist playing a very quick cadence, the speed of the movements is too fast for their organisation to depend on recurring stimuli from the muscles of the hand. If the speed of nervous conduction is taken into account, each movement will have to have left the brain before the stimuli born of the preceding movement has had time to return. The motor schemata must therefore be temporally organised in the cerebral cortex. Particularly quick instinctive behaviours also lead to the same conclusion. This observation of Lashley's appears to significantly depart from Lehrman's conception of the role of proprioception and to lend support to the thesis advanced by Tinbergen and Lorenz. In reality, this central, improvised and active organisation has nothing in common with the functioning of a nervous mechanism, or cerebral sections which would only have to be stimulated.[18] Even the pianist playing by heart actively organises the motor sequence, just like someone uttering a phrase or reciting something from heart actively organises its syntactic form. The goose that moves an egg back to her nest, the duck that oils its plumes or the fish that refresh the water in their nests by reversing the swimming motion of their pectoral fins are certainly more similar to a pianist who globally organises a motor sequence than to a mechanical piano that functions – as is shown by the phenomena of transport and instantaneous organisation, which are found as often in instinct as in habit. If the 'heart' of an instinctive act were a procedure linked to preformed nervous mechanism, it would not be able to adapt to circumstances by guide-stimuli given that such an adaptation often requires the general displacement of an act to another group of muscles and neurons. A musical rhythm is not imposed on its execution the way that the mechanism of a metronome can be imposed on a performance but is a rhythmic theme to which movements of performance adapt. 'If the leader of a quartet speeds up the time or retards, all the movements of the players change in rate accordingly [. . .] The violinist, in a passage requiring the whole bow, will draw the bow from frog to tip at a uniform rate for the required number of beats, whether the tempo is fast or slow'.[19] Von Holst demonstrates the same for a fish's swimming rhythm, where there is a thematic contamination of the rhythm of different fins, and even of respiration, which would be inexplicable if each of these rhythms had a local functioning.

All the misfires of instinct, the displacements and small accidents of substitution in the course of an act invoked by Tinbergen, are the equivalents of signifying lapses, written errors or grammatical mistakes rather than typographical errors resulting from a mechanical accident. No typing textbook would confuse a typographical error resulting from the proximity of one key and another on a keyboard with the fact that, in the mind of the typist, the 'plural' theme is contagious like a melodic rhythm and can incorrectly order another subordinated theme. Take the phrase 'He looked at them'. If it is wrongly retyped as 'He lookef at them', the substitution of the *f* for the *d* is clearly due to the proximity of two letters on the material keyboard. But the spelling error that produces 'He look at them' or 'He looked at thems' can hardly be attributed to the vicinity, in the human mind, of two groups of neurons of which one contains 'the movements of the fingers for the singular' and the other 'the movements of the fingers for the plural'.[20] The erroneous 'He look at them' is closely analogous to the 'local' development, in keeping with its own species, of a graft in an interspecies experiment. The general theme 'plural', stimulated by the pronoun 'them', induces the plural of the verb by contamination but according to the particular theme of a verbal, rather than nominal, plural.

Lapses of instinct, or lapses in language or human activity, may not always be meaningful in the Freudian sense, but they are always thematic. They are the result of psychic rather than mechanical or neurological accidents in the strict sense. With good reason, Tinbergen elaborates an analogy between the respective displacement activities in human beings and animals in an embarrassing situation. A bird, hesitating between fight and flight, straightens its feathers or even goes to sleep. An embarrassed woman readjusts her hair or touches up her lipstick. Before a fight, combatants can experience an overwhelming desire to sleep. But if this contrast is warranted, he rejects the strictly neurological interpretation of instinctive movements and displacement activities. It would be difficult to show that there exist, in the diencephalon of a human being, completely developed motor centres provoked by a diverted nervous influx from a subordinated centre to direct the motions 'use the plural', 'fix your hair' or 'touch up your lipstick'. It is difficult to believe that, were it possible to place electrodes at certain points of the diencephalon of humans (as we can do with cats), we would be able to trigger these behaviours.

There are a number of cases in which it is likely that certain peculiarities of behaviour are the result of mechanical accidents involving a pure proximity of nerves rather than psychic lapses. Accidents that induce epilepsy are of this kind. But nobody would confuse narcolepsy in the wake of encephalitis, or a traumatism, an irritation of the sleep centre which produces an epileptic crisis in the course of any activity whatsoever, for example, with the 'psychic' sleep in which the neurotic 'takes refuge' in sleep. Without question there is a nervous centre of sleep or, to be more precise, a centre of activation for the instinct to sleep. This centre can be stimulated just as well by the mechanical proximity of a centre of irritation as it can by a psychic theme. Equally, though, the key *s* on the keyboard of a typewriter can be pressed just as well as the result of a mechanical accident as it can as the result of the desire to denote the plural. A casual observer could perhaps denote the *s* key 'the centre of the plural' (and *a* and the set *es* the 'secondary centres of the plural').[21] But nobody can seriously assert that the cerebral motor zones contain in advance the movements composing the search for a lair or a quiet place to sleep, or the movements composing the search for food – any more than the keys of a machine do not contain in advance the varied functions which constitute the normal act of typing – in anything other than words alone. A more advanced automat than a typewriter is conceivable, one capable, through various feedback loops, of producing various well-formed phrases according to grammatical rules embodied in its assembly. The production of translation machines has also been attempted. But these auxiliary automatisms, which certainly find their own more advanced equivalents in the brain, are nothing other than more complex 'keys', instruments in the service of a non-localisable psychic theme and whose temporal procedures do not depend on any kind of reading or scanning* which would transpose a completely formed structure of melodic successions into a spatial structure.

It is a fact of experience that in the effort of recollection, we take ourselves to be capable of 'reading' or exploring an image or a spatialised schema in temporal succession. We can recite the months of the year backwards with the impression of parading an interior gaze over the mental schema of a calendar. We can repeat a short series of numbers out loud that we can take as having one meaning as much as any other. Reproductive memory almost always transposes an order which appears as spatial in itself into a temporal sequence. A physiologist is tempted to interpret this fact by saying that a part of the nervous system, equivalent to a television 'player', explores in succession another stable part of the nervous system. But even if this claim is admitted, it is clear that the difficulty has only been transposed. The reading or exploring part must itself be directly temporalised. Its temporalisation cannot yet be the result of a preceding 'reading'. It would be absurd, above all, to claim that when a semi-improvised action is in question – for instance, the pronunciation of a phrase or the performance of a non-habitual action – that the phrase or action existed first as a completely composed spatial structure in a corner of the brain and that another cerebral part, according to its own movement, encounters or clarifies it step by step, transforming it into a 'melody'. We do not 'read' our thought or our plans for action by speaking or acting: the action or phrase *forms itself*. A non-spatial and non-temporal intention becomes spatial and temporal first, passing through various increasingly structured states, which strikingly recall the study of aphasics. It is not first a spatial order and then a temporal order. Hess and Brügger's electrode, even in an animal's brain, can only act by awakening, indirectly, a thematic intention.

Physicians and even philosophers[22] are sometimes tempted to take the four-dimensional universe literally and seriously, in which, through a displacement of the line (or rather the zone) of time pre-existing events are encountered, the appearance of succession is produced. In psycho-biology, such an interpretation is excluded in advance. Instinct and morphogenesis are too difficult to separate, too indiscernible for any theory to be adopted for one that does not apply to the other. On the basis of all the evidence, what is already certainly false for a pronounced phrase or a meaningful activity is even more certainly so for morphological formation. There is no possible reading or scanning* in the formation of an embryo. Which part would be the reader, and which the read? The progress of differentiation takes place everywhere all at once. Lorenz and Tinbergen often insist on the fact that complex and distinct mechanisms exist in the central nervous system that can explain the autonomous procedures that they have discovered. They protest against the excess of theories of 'totality', which tend to make of the brain a kind of amorphous and homogenous jelly. It's fortunate, Lorenz says, that the organism has 'good hard particulate structures*' that perform specific functions.[23] But it is a profound mistake to believe that these structures are equivalent to an armed trap or a phonograph record and that their autonomous procedures are only their functioning. The egg, or the embryo in development, is not an 'amorphous jelly' either. But it is just as false to say that it possesses 'good hard particulate structures' such that the temporal procedures of formation will only be a 'reading' of pre-formed structures. It is beyond question that the temporal procedures of morphogenesis, like those of instinct, must be taken as truly elementary, as truly autonomous melodies, as the direct expression of themes which organise space and time *and which have no need to first exist in space in order to then exist in time.*

INTERPRETATION OF LEHRMAN'S THEORY

Let's now consider the position of the American school. Hebb and Lehrman cannot be reproached for being preformationists. They even seem to have succumbed to the inverse excess, effectively believing that almost everything in instinct is improvised, discovered through good luck and then perfected through apprenticeship. They even have the tendency to see apprenticeship, adaptation through trial and error, at work not just in instinct but also in formation, interpreting every epigenetic factor as the index of learning*. As there is with every exaggeration, there is something troubling about this exaggeration of the notion of learning*. Something rings false in Lehrman's description of the instinct of the pigeon or rat, despite the facts and the recounted experiments, a false note that is easy to detect.

If instinct is composed through a fortunate sequence of apprenticeships; if the rat learns to make her nest by first moving around objects in general and learns to clean rather than eat her young by first cleaning her own genitals; if, furthermore, pigeons learn by chance to feed their young by cleaning the feathers on their own chest; if, as Kuo maintains, cats must learn to chase mice;[24] and even if, as Kuo and Lehrman sometimes seem to claim, chicks learn how to peck by having their heads passively moved up and down by the beating of their hearts, we can still ask by what miracle the initial moment in these chains nevertheless almost always leads to a typical and effective behaviour, in nature if not in the laboratory. If the organisation of instinct, and even morphogenesis, has a 'history' – in the precise sense of the word, that is, a sequence of chance events organised and made use of more successfully than unsuccessfully – why is it that the biological history of an individual gives rise to strictly specific forms rather than forms organised randomly? Cournot once said of human history that when chance seems to continually bring enterprises to ruin, this is proof that they are not ruined by chance.[25] If we replace 'bring to ruin' with 'lead to succeed', and 'human history' with 'the history of the animal', the conclusion is the same. *If chance seems to always lead an animal to succeed, this is the best proof that it does not succeed by chance.* Hebb and Lehrman would perhaps protest against the word 'chance', which they carefully avoid using. They only speak of adaptations to the exterior world, to internal or external milieu and to transfers of learning*. But chance, and even miraculous chance, must indeed be in play in order for these adaptations to circumstance to discover at the same time appropriate adaptations to *vital necessities*. It is clear that if the stimuli given by the milieu, if the various irritations and tensions provoked by hormones are only 'stimulus-drives' – to which the animal responds as it can by improvising its response until it rids itself of the irritation and diminishes the tension – then an incredible series of chances that lead to a specific type must have occurred. This alleged anti-finalist conception requires, to the contrary, a marvellous providence.

If both external and internal are, for the animal, not drives* but discontinuous signals to which it knows how to respond in advance – with, in short, the obligation to engage in pseudo-learning*, which is in reality nothing other than an adjustment – then, to the contrary, we understand very well the specificity of response. To recall Mandelbrot's analysis, the animal is equivalent to a signal decoder armed with a code which allows it to not only repeat but also to recover the text. This is why each individual can reproduce the 'text' of the forms and behaviours of its species exactly *despite*, rather than *thanks to*, the chances that affect it as an individual. It is only under these circumstances that the 'message' of the milieu to the individual – due to the specific code possessed by the individual as a 'competence', which is perhaps, according to Mandelbrot's expression, 'a-chronic' – does not change with time. It is only under these circumstances that the development of the individual may not resemble an historical event.

LEARNT BEHAVIOUR AND NORMALITY

Experts in the functionalist theory of morphogenesis or instinct as the continuous function of a milieu often fearlessly advance it before finally accepting its logical consequences, namely, the negation not only of every idea of type or species but of every idea of normality: 'Thus we cannot even speak of certain structural characteristics as being "normal" for a given species and fixed by hereditary constitution. If the environment in which the organisms develop were to undergo a change of a more or less permanent nature, a different set of characteristics would come to be considered normal'.[26] When cold temperatures or insufficient oxygenation, ultraviolet radiation or chemical substances like lithium and magnesium act on the embryos of sea urchins, fish or *Anura*,[27] producing monstrous attached doubles, cyclopeans or individuals with skeletal excess, it is necessary, in light of theory, to put the word 'monsters' or 'excess' in quotation marks. The sick human beings to which G. Canguilhem rightly makes reference in developing a philosophy of 'normality' would certainly see things differently. Doubtless, if the new 'norm' implies the impossibility of surviving – if, for example, pigeons lacking the stimulus do not learn how to feed their young – then the species will disappear through natural selection. But if selection can explain that the starting point is a form or effective behaviour, it cannot explain why that form or behaviour is typical. Pigeons have no trouble surviving by drinking the way other birds do. And to respond by saying that selection acts to maintain those genes which determine a form or a typical behaviour is to say nothing about, for instance, the polymorphisms of social insects that in themselves lack the identity of genes and for whom agents in the social milieu determine discontinuous forms, each of which can be called 'normal' and at once effective and specific. And there also exists a sort of polymorphism of instinctive responses. Grassé has shown that instinctive response, in circumstances that only slightly differ, can take the form of two or three markedly distinct and yet equally well-adapted types. What appears to be an improvised modification of instinct is thus often a kind of phenotype of replacement. Thus, during social division, normally nocturnal termites remain in the light; the king and queen walk instead of flying, displaying no sexual activity, and continue to transport eggs. There is therefore no improvised obedience to 'stimulus-drives' in instinctive and morphological responses but a codified reaction in accordance with distinct norms each of which is meaningful rather than random. Intermediary behaviours are the true anomalies, as is the case with all intersexuated individuals. Monstrosities are the result of the fact that the organism, placed in an artificial milieu, cannot, despite clearly observable attempts towards what is normal, correct the milieu. The sick human being calls the doctor, buys the prescribed medicine, takes a cure at a spa and, in short, sets out to escape the milieu which has 'denormalised' him. And on this point, animals are no different from human beings. Through their behaviour if not their techniques, they look for a way to escape their denormalisation, evading harmful situations by creating within them a zone of security or even, if Hédiger is to be believed, by searching for curative sustenance.[28]

What is of course clearly false with respect to the individual organic development of a human being is, on the contrary, probably true for the social development of, if not animals, then at least humans. The potentialities of development for an egg of a given species are almost precisely fixed, allowing us to speak of a normal, completed development. At the level of human social behaviour, as ethnography has clearly shown, this is no longer the case. No one type of cultural development can be said with certitude to be more normal or better than any other; all that can be said is that they are different. The potentialities of behavioural development – at least in human beings – are not specifically determined. All human post-natal development has an historical character – so long as what is true of human cultural development is not applied to organic development. That ethnologists have ceased speaking of primitive, inferior or abnormal cultures relative to Western culture does not mean that biologists should refrain from speaking of interruptions, deviations or anomalies of development.

THE CHARACTER OF INSTINCTIVE THEMATISM

The advantage of the American school's conception, is, we might say, that it is not preformationist. But on closer examination, the poverty of this advantage begins to show. The more instinct is taken to be an improvised regulation, the more it is necessary to presuppose entirely ready-made organs at the level of the morphology and physiology of the animal. If the rat has no need for a particular instinct when it comes to absorbing the requisite amount of salt, or the pigeon when it comes to feeding her young, then it becomes all the more essential, in order to understand these regulations, to begin with ready-made organic apparatuses. Lehrman's conception becomes, consequently, as inadequate as Lorenz's for understanding the intimate unity between instinct and morphogenesis and the manner in which the species 'rat' has been able to progressively exhibit the apparatuses which today assure, for instance, alimentary regulation. And if the domain of apprenticeship is expanded to include embryonic formation, it leads to still more paradoxical theses than it does with respect to instinct strictly speaking. If it is supposed to explain the passage from the pigeon egg to the adult pigeon with its 'milk'-producing crop in the same way that it explains the passage from the inflated crop to 'feed the young' behaviour, it is led to invoke even more miraculous chances.

We can thus conclude that both interpretations fail. The facts remain: there are autonomous or quasi-autonomous procedures, and there are semi-improvised behaviours, imperfect regulations that the animal must develop. Neither can be understood if they are linked to mechanical functioning, whether that of a cerebral centre, like the German school, or the whole organism considered as a homeostat in relation to its milieu, like the American school. To be understood, the action of a non-spatial thematism – which can take the form of either a mnemic melody in the case of autonomous procedures or a vague ideal, but one sufficient to orient and lead towards a 'normal' trial-and-error behaviour which appears to only have the immediate result of diminishing a tension or displace an irritant – must be recognised in both cases.

Consider the case of sexual instinct, in which stereotypical melodic fragments and specific releasers* are found.[29] A great deal of improvisation, and efforts to diminish to some degree the chance of peripheral tensions – not only in courting behaviour but also in the act of consummation – are also found there. But only the most obtuse of philistines would reduce instinct to a detumescence, even one regulated by proprioceptive sensations. Beyond mechanical or physiological feedback*, there is what could be called axiological feedback* in which the directing 'ideal' is not a form given in advance or inscribed in the organic machine but a kind of obscure phantom that grows clearer to the extent that the first behaviour draws near. For both human beings and animals, this directive ideal is the equivalent of a forgotten word which awaits not on

166 the threshold of recollection but the threshold of recognition. Luck and chance play in the interval between these two thresholds alone, as is often the case for instinct as well as for invention. Most inventors are lucky, but their true merit is to have recognised this passage. The animal 'recognises', through the satisfaction that confirms it, the conformity of its act with a norm or an ideal which must be described as the ideal of the species. The progress of its behaviour is not independent of luck but is subtly sanctioned and oriented by it.

As we have seen, and as E. S. Russell has shown, orientation primarily operates through a 'valenciation' of certain situations or objects, which thus become troubling, or fascinating, or attractive – important or interesting, in any case – and in the presence of which the living being, human or animal, experiences premonitions. Consequently, in the finalised conscious act, it is the vague image of the goal to attain which orients action; in the instinctive act, it is the valorised or 'valenced' object which plays this role; and the displacements of valenciation, to the extent that action takes place, guides the living being like Ariadne's thread.

Incidentally, it is often the case that even in the finalised conscious act, a valence projected onto an object-means is encountered as if cut off from the consciousness of the goal, the movement proceeds deceptively and the thread is broken. In this situation, the human being or higher animal resembles a bird or insect that embarks on a migration. Even human beings sometimes express their love through blows [*coups*].

Köhler provides an amusing example of this in his celebrated book. The monkey Coco has to resolve a problem 'in stages': he has to use a stick to get the bait but has to first use a crate to get the stick. The animal grasps the principle of the solution but during its execution, he forgets the exact role of the crate. And yet, the case remains 'valorised' or 'valenced'. Everything about the monkey's overall behaviour seems to cry out, 'I'm supposed to do something with this crate!' And yet, given his predicament, he beats on the crate with fury.

167 It would be difficult to explain this 'displacement activity' by noting that, as Tinbergen does with respect to instinct, there is a central tension or hesitation between two completely established nervous mechanisms since the animal is in the process of resolving an improvised problem. It would be just as difficult to say, with Hebb and Lehrman, that he is groping around on the basis of proprioceptive sensations. The 'irritant' is a visible object and not an engorged organ, and it is an irritant because it is 'valenced'. The inevitable consequence of a successfully accomplished instinctive act is not only pleasure or a local relaxation but satisfaction and the fulfilment of an aspiration. Need, experienced through proprioception, and desire for an object or a valorised being, are closely related and will meet up when the act succeeds.

On the whole, the descriptions of the American school seem more accurate for instinct than do descriptions involving 'autonomous procedures'. It is quite true that most often, the 'innate' element of an action is in some sense further *upstream* than the German school admits. The innate is held to act in its motricity so that what is taken as innate is the proprioceptive sensation which stimulates the act as an irritant or even, even further upstream, it is a hormonal secretion which provokes the proprioceptive sensation and which orients action in turn. Regardless of how, an act results in which the animal is engaged; but this state also bears on the valenciations which are 'before' and 'downstream' from the act and which explain its direction and its successful arrival. In sum, the motricity of the act itself is rarely innate. What is innate is, on the one hand, an 'irritant' which guides the act by driving it [*poussée*]; on the other, it is a valenciation set upon [*posée*] an object which guides the act through attraction. But in reality, the two elements of irritation and valenciation are often difficult to distinguish and are, furthermore, often confused with an innate motricity in the act that they 'enframe'. For that matter, all spatial metaphors are here misleading because the directive theme is outside space and dominates acts within space.

Regarding morphogenesis in the strict sense, it is the descriptions of the German school which appear most adequate – for morphogenesis, and also for certain instincts very close to morphogenesis, such as the instinct to suckle the newborn. Piaget was able to show that reflexive sucking movements appear before they have an object and that they do not adapt to it. He has shown that the object is first food for the activity of sucking before being food for the general instinct for nutrition. But the movements are not local functions. The living being is not the passive locus of movements which unfold within it – it executes them, it is the agent and not the machine-like support; it invents them along the way. False invention, or rather facilitated invention, in the sense that the organism is possessed *by* the theme, always acting *according to* the theme that it bears out.

An interpretation on the basis of trans-spatial thematism succeeds on both fronts by allowing for the rapprochement of both interpretations, *which are only incompatible if they are mechanist*. At the same time, this analysis of instinct allows us to understand what truly constitutes morphogenetic competence: a kind of non-proprioceptive habit, unregulated by perception – since the developing embryo does not have sensations in the strict sense, and still less the possibility of seeing valorised objects – an 'autoproprioceptive' habit, if one can hazard this somewhat atrocious term, which is to say, one guided by its own tendency to continue according to a dominant theme. To recite a well-known text by heart, this union between activity and passivity that is so difficult to explain is directly demonstrated. Every accidental break is felt to contradict an internal dynamism towards a defined structuration, belonging at once to the recited text and the reciting self – a dynamism not at root composed of a localisable need, like that of a thirst prevented from being quenched, a dynamism that would be impossible to imitate through the assembly of a homeostat. An automated homeostat could be designed to be capable not, of course, of experiencing thirst as a lack but of drinking and even trying to continue drinking despite interruptions, where recurrent indications emanating from an indexed reservoir would incite movements of absorption in it. A non-homeostatic automat could also be designed to recite a fable by La Fontaine thanks to the unspooling of a perforated band directing an artificial larynx. But what could not be designed is a homeostatic automat capable of reciting a fable. Outside of the organism, simple dynamism (through equilibrium) and complex mechanism (through piece-by-piece adjustment) can only be encountered separately. Only life is capable of uniting, in morphogenesis, dynamism and complexity. Dynamism towards a complex structuration and not a simple equilibrium, an 'auto-proprioceptive' melody, guidance by valenciation or axiological feedback* towards a non-materialisable idea – all of these phenomenon are extremely close to each other and characterise both instinctive competence and morphogenetic competence to the same degree.

Chapter 8

Open Formations and Markovian Jargon

The particular appearance of all of the not strictly typical organic formations – in which formative instinct is at the mercy of chance, which it cannot succeed in dominating, and in which, consequently, something historical appears (in Cournot's sense, if not Aron's) – remains to be studied. When the animal manages to live in an *Umwelt* almost as standardised as the egg or the maternal womb, it can produce forms of behaviour or of extra-organic construction almost as typical as the forms of its organism, and according to analogous procedures: autonomous melodies, and feedback* by valenciation. Every living being seeks to escape from the accidental, from history, and in fact succeeds, under the pain of death, in the most decisive phase, the embryonic phase of its development, which always takes place in a sheltered zone[1] and, we might say, ideally tranquil and Arcadian conditions, even if after its birth, it is destined to fall into a world stalked by cataclysm. After birth and save some misfortune, the young of the higher species continue to be sheltered by the familial and social milieu. But it will unavoidably encounter adventures, and when it becomes an adult it will never entirely succeed in dominating chance, in either life or production. Autonomous melodies and regulations are overwhelmed; produced forms are, in turn, mixtures of the organic and the fortuitous; they have the appearance of 'open', unsecured or poorly enclosed forms in which unexpected stimuli have provoked responses which are not in themselves devoid of sense but whose sense is disjointed, and which call in their turn for other responses, leading to adventure.

MARKOV CHAINS

In order to define this mixture, we will present a first approximation in the formal schemas in which mathematicians have constructed mixtures of *alea* and dependence. Consider a series of successive draws D1, D2, Dn from a lottery where there is a total independence of each draw. Now, consider a law of dependence according to which the probability of Dn depends on Dn-1, on the set {Dn-1, Dn-2} or on some more complex set selected from the anterior draws. We now have a Markov chain, named after the Russian mathematician who first studied partially dependent aleatory phenomena.

A concrete example can be drawn from the very domain in which Markov himself first applied his models: a machine that automatically produces a pastiche of a language.[2] For any given language, constant and characteristic statistical values exist for the use of certain letters and groups of letters, for the mean length of words or series of letters or even for the use of words and series of words.[3] In French, for example, the letter q is always followed by u, h is preceded by c in almost half of the cases, while e is followed by another e less frequently than it is in English. If, for example, draws are taken from a series of boxes containing, according to their frequency of use, groups of three letters, the boxes are arranged in the alphabetical order of the first two letters of triads they contain and it is agreed that after the first draw the second will be taken from the box indicated by the last two letters of the drawn triad, in this way a sort of jargon will end up being produced, one with the appearance of language whose mean frequencies and sequences have been established. Guilbaud gives the results of such a draw organised according to the frequency of triads in Titus Livius:[4]

IBUS CENT IPITIA VETIS IPSE CUM

VIVIUS SE ACETITI DEDENTUR

IBU would have been drawn first; then, from the BU box, BUS; then, from the US box, USC would have been drawn; then, from the SC box, SCE would have been taken, etc. The resulting phrase, a Markov chain, unquestionably has the air of being Latin, just as it would have the air of being English if the boxes had been populated according to the frequencies and mean sequences of English. Authors of pastiche instinctively conform to the typical frequency of words or turns of phrase of their victims. A pastiche of Baudelaire could not but use the words *ange, parfum, sein, extase, sang* and *démon*.[5] For Mallarmé, one statistical study has revealed the words used more frequently than they are in current language use would be *azur, nue, vierge, or, rêve* and *pur*.[6]

BIOLOGICAL 'JARGON'

An animal, let's note, will easily allow itself to take up a parody of the same kind, drawn from the signal stimuli that interest it and originating with its peers in the milieu. In order to approach ostriches without arousing their suspicion, indigenous ostrich hunters will not simply disguise themselves as ostriches; they will mimic the modes and sequences of the animal's behaviour. The sparrowhawk's hunters are forced to give their lures the typical behaviour of a sparrowhawk landing. Animal behaviour is often, in fact, a semi-fortuitous enchaining of themes evoked by the preceding phase and lacking an overall plan. This is not a language but a jargon. And if the animal so easily confuses language and jargon, it is because it is, itself, 'jargoned'.

An autonomous melody, a regulation in normal development or in the strict activity of important instincts does not resemble a Markov chain. Each stage of its unfolding does not depend on the preceding phase alone nor the sum or totality of all the preceding phases. It is 'surveyed' by a dominant theme, like the unfolding of a well-organised phase, in which lapses are compensated for and do not lead to aberrant divergences. If things were otherwise, a bird, for example, would never be able to feed its young by carrying a worm in its beak without swallowing it. By contrast, in the freer and, necessarily, more improvised and historical regions of instinct – and even in the development of plants, where development takes on the appearance, to use Plessner's expression, of a *Gewächs*[7] – we are very close to the Markovian schema.

MARKOV CHAINS AND THEMATIC ORDER

It is necessary to insist that even in a completely Markovian biological chain, vertical thematism is not absent. The possibility of constructing a machine that will itself effectuate the selection of draws and the 'speaking' of jargon – like A. Ducrocq's Calliope[8] – might mislead us into believing that there is a pure mechanism in everything that is Markovian. But we must not forget that in the example discussed above, the constitution of the triads is inspired by the phonetic character of language. Although each draw is mechanical or mechanisable in its execution, it bears on a theme, and, furthermore, the passage from one box to the next also takes place in accordance with a theme borrowed from language. It is even clearer that each segment of a Markov chain of instinctive behaviour is thematic. The schemata of development and of normal behaviour, in which the final part of the first section evokes the subsequent section, are always applied, with the difference that these sequences are not 'enclosed'. In psycho-biology, the sequencing of one theme to the next does not happen by drawing from a box but, what comes to the same thing, on the basis of the constellation of stimuli that evokes, in the mnemic sense of the word, the subsequent theme. Or rather, it is the draw from the box that resembles mnemic evocation. In effect, the man drawing from the box is obliged to read the final group of letters in order to guide the choice of the subsequent box: he lends his consciousness to the operation. Even if he shows the group of letters to a machine capable of mechanically reading them and thereby triggers the subsequent draw, this assembly, which will make use of a system of keys, clearly depends on a first psychological reading and will only be a materialisation second. In other words, the assembly of a Markov chain would not be able to do without an elementary consciousness, analogous to that of Maxwell's demon.

And this is why, incidentally, a Markov jargon, which is completely disorganised relative to a normal discourse unified by a dominant theme, is relatively unified by a non-sequenced, or less sequenced, chance draw. It is midway between order and disorder, more or less ordered depending on the themes enchained, which themselves contain more or less order, or a higher or lower degree of order. From a machine for constructing phonetic pseudo-French we can pass to a machine for constructing morphological pseudo-French, and then syntactic pseudo-French, and then, for example, to pseudo-Baudelaire. A machine can even be designed to display pseudo-insect behaviour, and then bee behaviour, Italian bee behaviour, etc. But, properly understood, this does not allow us to conclude that the passage from disorder to order is the progressive result of pure mechanism since the engineer contributes order to the successive improvements of the machine, inspired by the order that is already realised in a language or a specific behaviour. Jargon always presupposes a language, clearly, and it would be absurd to claim that language is constituted on the basis of jargon. Chance is like a share house [*Le hasard est comme l'auberge espagnole*] – it can only order what is brought into it and that it contains. The proverbial typing monkey would clearly have a greater chance of reconstituting a verse or two of Baudelaire if it were typing on a machine pre-arranged to produce Baudelarian jargon rather than one arranged to produce pseudo-French, and an even greater chance again than if it were using an ordinary machine – but it would be bold to conclude that we are close to a mechanical explanation for Baudelarian poetry. This would be putting things backwards – such that disorder is produced on the basis of order – to assert that the biological or linguistic register can be attained by explaining them on the basis of the disorder of pure draws progressively modified by increasingly thematised Markovian draws.

THE STATISTICS OF BEHAVIOUR

If we examine, as the theoreticians of information (Zipf, Shannon, Mandelbrot) have done, the statistics of words in a statement or a sufficiently long piece of work, we find evidence for strangely simple general laws. After Zipf, for instance, we can assert a very simple relation between the frequency F of a word in a given text and its rank R, organised in order of frequency: R x F = a constant.⁹

For example, in J. Joyce's novel *Ulysses*, the word of rank 10 is used 2,653 times = 26,530

100 x 265 = 26,500
1,000 x 26 = 26,000
10,000 x 2 = 20,000
29,000 x 1 = 29,000

Shannon and Mandelbrot have shown that we can get closer to the facts by substituting a somewhat more complicated formula for Zipf's law, but the principle is the same: it is possible to treat a text in the same way that we treat the characteristics of a gas: statistically, which is to say in ignorance of the particular causes that microscopically determine the positions of molecules and words. The 'temperature' of a text can thus be defined, or the slope of the rank-frequency curve, where an 'elevated temperature' signifies that the available words are properly used and that rare words are being used with significant frequency.

> The same holds in an apparently different domain where, by photographing a fly flying under a screen at regular intervals and then correlating the photographs' points, P. Vendryès has obtained a trajectory analogous to that which is obtained by photographing a particle animated by Brownian motion in a liquid, participating in chance molecular agitation.[10] These same trajectories, which are, if not exactly Brownian and aleatory, at least 'Brownoid',[11] are also obtained when the movements of leukocytes are filmed, and even the positions of a Parisian cab if noted at regular intervals during the course of a working day. These trajectories are not exactly Brownian. In a supposedly indefinite fluid, the probable distance of a particle P from its point of origin is equal to its free mean route multiplied by the square root of the number of routes; this distance therefore regularly grows, on average, with the square root of time, something that obviously cannot be the case for the fly, the leukocyte or the cab. It would be more accurate to consider these as Markovian, or Markovoid. Each taxi or each stimulus which influences the leukocyte or the fly is like the draw from an urn containing, according to their frequency, its behavioural themes. The journeys asked of the cab driver by his passengers play the same role, the position of the car, having been acquired from its preceding course, orienting the request for the next journey. Each course is clearly thematic and planned, but the succession of courses, without being the result of pure chance – since each new course depends in part on the preceding course – is sufficiently aleatory to be non-thematic and to resemble an effect of pure chance.

Whether division is Brownoid or Markovoid, what matters is that as soon as there is a division of themes and the intervention of chance in their enchaining, the nature of themes no longer plays a role and appears to be eliminated. It is in this way that the possible rapprochement of an ordered text and a disordered gas (that is, one ordered by chance alone), if not exactly Zipf's law, can be understood. Each of Joyce's phrases is obviously thematic, and the same holds for each paragraph – it is a question of language, not jargon. But when a large number of phrases are considered, it no longer matters whether the whole is thematic or not, and general and statistical laws prevail. The directions of particular themes are neutralised by their great number. Zipf himself interpreted his law as the result of an equilibrium between two forces in discourse: on the one hand, laziness, and the exercise of the least amount of effort on the part of the speaker, who uses the fewest possible words and would choose to call everything a 'doodad' or 'thing'; on the other, the need to express each idea clearly by sufficiently structuring it and sufficiently informing the listener. If the discourse is sufficiently long, the particular character of each theme can be abstracted from it in order to express only what remains of the general force of structuration. Doubtless when Joyce wrote *Ulysses*, he wrote a novel which had a subject and a central theme, but it would be even truer to say that he 'did a Joyce', just as the leukocyte 'does the behaviour of the leukocyte' or the fly 'does the behaviour of the fly'. The characteristics of form are found much more in the texture than in the organisation of the whole. Balzac only became aware of the unity of the *Human Comedy* retrospectively. He essentially 'did a Balzac'; in turn, a statistician could not distinguish, on the basis of a word count, between novels by Joyce and Balzac, between a Balzacian or a Joycean jargon constituted on the basis of Markov chains. If, on the contrary, the statistician considers an isolated paragraph from Balzac, he will have a much smaller chance of finding it typically Balzacian from the point of view of vocabulary than he would if he considered a random section of Balzacian jargon of the same length, which is only morphological in its texture. It is notable, to recall a remark made by Mandelbrot, that if the texts of normal subjects are being studied, it is necessary to take a great number of precautions to avoid distorting – if it can be put this way – short samples given that every text is devoted to a particular topic, whereas the language of schizophrenics, on the contrary, whose mechanical capacity to emit words is intact even though the correspondence between words and ideas is profoundly affected, much more regularly confirms Zipf's law even in very short samples.[12]

The behaviour or language of animals and normal human beings allows for a return to statistical laws if they are considered in sufficient quantity. Only a superhuman could organise his whole life in such a way to make no false gesture or pronounce a single word which was not formally attached to a unique design, and for whom there would be no jargon but only a pure language. What Balzac, Joyce or Proust, despite their efforts and declarations, which are not free of boasting, could not truly succeed in doing in a novel is even more impossible to do in real life.

CULTURES AS MARKOVIAN KEYBOARDS

And what is impossible for a human life is even further from being possible for the life of a group of human beings or for humanity in its entirety – outside of the works of Hegel.

Certain cultures belonging to a circumscribed time and space are, without of course being analogous to a meaningful language, at least analogous to a strongly thematised Markovian jargon, or to a set of boxes prepared for such a jargon. They are stylised and possess a behavioural vocabulary and syntax, restrained and closed in the fashion of a tragedy by Racine or a poem like *Narcissus* by Valéry, about which the enumeration of words reveals an abnormally and voluntarily limited vocabulary. In this kind of culture, it is often the case that the most ordinary people easily participate in the collective style in the same way that the least of the Greek lyric poets of the Homeric school easily 'writes a Homer' to such a significant degree that the Markovian instrument provided by tradition only needs to provide a minimal surplus of order for the level of the work of art to be attained.

Like every language, every culture is, to varying degrees, analogous to a Markovian keyboard offered to each individual. If the individual were to improvise everything, they would remain in infancy. But in fact, thanks to the strong thematisation of culture, its contribution can be limited to the adjunction of a last coordinating theme; it easily attains the level of a behaviour that is at once stylised according to culture and oriented by personal goals. 'Pueblo ritual', E. C. Parsons writes, 'is kaleidoscopic. There are numerous types of rituals, and they are combined in diverse ways. With or without a dramatic theme, these rites constitute a ceremony . . . The rites are combined and recombined; the rite itself is fixed and conventional, but the combinations are less rigid: in fact, the suppleness of Pueblo ritual is striking'.[13]

Sages, of which Confucius is the exemplar, have clearly realised the necessity of not asking too much of human beings. They wish for society to provide a complete keyboard to each person, along with an assortment of forms and rites, at the risk of making human life resemble that of an animal only able to draw upon the ritualised themes of specific instinct.

It is clear that if the collection of individuals who participate in even strongly stylised cultures is considered, the set of their acts, like the set of phrases that they pronounce, can only be treated in a statistical fashion and that they resemble a Markovian 'tissue'. This set can be characterised, as R. Benedict and Kardiner have done, if not without some arbitrariness, by a certain dominant style – whether religious, warlike, competitive – just as it can be said, somewhat arbitrarily, that a certain language is particularly suited for juridical expression, for giving directions or for everyday conversation. But it would be absurd to say that a certain culture signifies a certain idea, just as it would be to say that French or German has a meaning. Only a phrase pronounced by an individual can have a meaning.

This is even more certainly the case if the totality of human cultures and human history, strictly speaking, above all political history, are taken into account. They resemble a chain, or rather a Markov tissue, much more than they do the Hegelian dialectic. It is a jargon, not the expression of a *Logos*, and the interpreter who stubbornly searches for its precise meaning is the equivalent of a vain novitiate in Latin who could be pranked by giving him Guilband's pseudo–Titus Livius to translate but who would claim that he had discovered its meaning.

What complicates the problem in the case of human history is the fact that it is not homogeneous. Political history has a Markovian appearance. From the political point of view, the unfolding of a great revolutionary crisis closely resembles a sequence of random draws of thematic actions. Consciousness and human illusions only trouble the unfolding in the vain effort to master them, and sometimes by complicating the involuntary jargon through voluntary parody: the French revolutionaries were playing at being Romans, and Bonaparte in Egypt copied Caesar in Gaul. Cultural history also appears to be Markovian, if not as dramatically. Semi-fortuitous sequences can be found there, above all in the passages from one set of values to another, where one finds, as Max Weber says, a 'prodigious interlacing of reciprocal influences'.[14] With respect to a particular event, the historian or the sociologist thus searches for what they call the new form – by virtue of neither causality strictly speaking nor pure logic but the resonance of themes or a partial elective affinity in one that is already present, as in a Markov chain, or rather in the mind of the operator who constitutes it according to conventions, BUS following IBU – in the forms that are already present. The classic example, drawn from Max Weber, is the 'calling' of the 'capitalism' form by the 'Calvinism' form, or again, in a slightly different sense, the extension of wine culture into Christian countries through the ritual obligation of taking wine in mass, or the domestication of the chicken for divination. History is not a dialectic. Merleau-Ponty puts it very well: 'History includes dialectical facts and nascent significations; it is not a coherent system. Like a distracted interlocutor, it allows the debate to become side-tracked; it forgets the givens of the problem along the way'.[15] It is only dialectical if the word is used with an accent on its etymology, or the sequenced or re-sequenced divisions, on the passage and not the unity of its *Logos*.

While social evolution and even more so technical evolution do not escape from Markov chains, they do also obey different laws. Technical progress most closely resembles – at least if the point of view of a limited technique is adopted – a unitary morphogenesis of a biological species. With respect to technical progress as a whole – as F. Meyer has recently remarked[16] – it can be represented as the filling-in of the envelope of a curve with a 'slope' even steeper than that of an exponential curve. What is obviously excluded from technical progress, taken as a whole, is its Markovian character. The steep progress of human technique appears instead to be explicable by analogy with the sequence of factorials. The two are related by the fact that each new acquisition is not merely added to those that preceded it but multiplies them, as Leibniz said of the psychological faculties. The technical 'morphogenesis' of the automobile presupposes the preliminary deployment of numerous techniques – metallurgy, electricity, the extraction of petrol, the harvesting of rubber, etc. – without even considering the social organisation of communication channels. A scientist or engineer of the seventeenth century would have absolutely no chance of creating an automobile anything like our actual vehicles – he would have had to possess a multivalent and superhuman degree of genius and, at the same time, have had to create a multitude of components and indispensable techniques. The engineers of the nineteenth century, in contrast, would have only had to carry out a number of 'multiplications'. The inventors of the aeroplane had, in turn, only to combine the combustion engine and the glider. Jet aircraft were developed much more quickly than those with piston engines. Technical 'elements' like the screw, the universal joint, the magnet, the photoelectric cell or the three-electrode lamp, themselves very complex, lend themselves to more and more varied combinations; they are deployed in products whose general curve therefore corresponds, broadly speaking, to the progression of factorials.

Economic life also displays a mixture of the aleatory and the thematic, but in a way that is not in general Markovian. Modern economic theory reacts to classical economics – which passes from the microscopic, that is, from the individual behaviour of *homo oeconomicus*, to the macroscopic, that is, to the social economy, only considering the statistical effects of individual behaviour and likened human society, in its life or rather its economic equilibrium, with a pure physical system in which everything is quantitative – by instead likening society to an imperfect organism in which themes or finalities, which are super-individual even though they emerge, through the intermediary of heterogenous social groups, from individual themes, are combined with the purely statistical equilibria that they enframe. Society is a hierarchy of these psycho-social forms and not a kind of homogenous milieu. These psycho-social groups are capable of both anticipatory drives and of forming habits which postpone decisions, of both strategic calculation and mental inertia and the association of ideas.[17] Human society, in its economic life, is neither a physical system whose evolution would only be a march towards equilibrium nor an organism whose development is entirely thematised and predictable. Economic life has a truly historical character even though it does not appear to be Markovian. It resembles, rather, a strategy of the kind schematised by von Neumann and Morgenstein which, while timeless as a pure schema, becomes historical from the moment that the game is played by social groups whose conduct is not entirely rationalisable.[18]

LIVING SPECIES AS MARKOVIAN SYSTEMS

If we return to biology now, and consider the history of living species rather than that of human societies, the Markovian schema has an equally large scope of application. The palaeontologist who works to closely follow all of the twists and turns of the evolution of a species or an order and, even more so, he who works to turn his thought towards the morphogenesis of particular flora or fauna, has a very different experience than that of the biologist who studies development and individual morphogenesis. For the latter, everything is order and reason. Nothing resembles the disorder of jargon or the disorder of history less than the admirably effective and 'timed' unfolding, free of hesitation and chance, by which, over the course of months or even weeks, an imperceptible cell makes a marvellous animal. But if the palaeontological history that leads to the ant, the giraffe or the human being were or could be followed, it would convey a completely different impression. A history of England or of Italy would appear to be simplicity itself in comparison to the adventures and avatars, contingencies and crises, revolutions and realignments which have made the passage from primitive living beings to an insect or an actual vertebrate possible. It is at this point that contemporary biologists are almost as sceptical about the possibility, and even the meaning, of a theory of evolution whose pretension is to provide a universal key for understanding everything, as professional historians are about the possibility of a philosophy of history.

MARKOV CHAINS IN EVOLUTION

Without the concept of 'open formations', it would be impossible to understand biological crowd phenomena, which display the trace of semi-fortuitous sequences even more clearly, of which there can only be an historical explanation. Parasitism, symbiosis or commensalism,[19] which play such a large role in the constitution of species, are striking examples. The morphology of a crowd of species is 'synthetic', the crowd seemingly composed of themes gathered together by chance, as if typed on a Markovian keyboard, and only then organised into more logical phrases.

Many animals function as the organs of other organisms and are incorporated and transmitted in their reproduction. Cuttlefish function as the filtering organs for luminous bacteria. The worm *Microstomum linearum* is equipped with stinging cells, but Kepner has shown that these weapons are taken from the microscopic hydrae that the worm feeds on. When the hydrae are digested, the amiboid cells that line the worm's stomach gather the stinging cells and transport them to the mesoderm, where other mobile cells gather and transport them to the skin, turning them into a sort of poison-bearing tube oriented towards the exterior.[20] The Brazilian fungus-growing ants studied by Autuori[21] are veritable symbiotes of the fungus which they sow, cultivate and reap; which the female carries on her nuptial flight in an intra-buccal pocket; and which are encountered nowhere outside of the nests of the ant *Atta*.

Slave ants are more interesting yet again here since they constitute a perfect example of the Markovian origin of a specific trait. Most species of ants spontaneously attack ants of different species and carry away plunder if they are victorious. The most precious plunder – at once because of its habitual association with 'precious object' behaviour in its own anthill, because of the hostile effort expended to conserve it and because of the superior nourishment it provides – is the brood, eggs and larvae. The vanquishers devour the eggs upon their return.[22] But it is at this point that the pupae begin to hatch, eventually helped by the first blows of the predator's mandibles.

The meal thus immediately stops since the ants that then hatch awaken the parental care instinct instead of the instinct for nutrition or combat. No longer enemies or food, they are indistinguishable for the vanquishers from their own young, and, in turn, the young take their abductors to be their parents. The various effects in different species of this first 'derailing' can be followed in a Markov chain as far as the extreme case of *Tetramorium atratulum* where the predators are no longer capable of producing workers and must replace the queens of an exploited species with their own queens.[23]

Hegelian and Platonic interpretations of such specific formations appear equally poor, while the analogy with linguistic evolution is excellent – according to which they are produced by the faults of language, themselves due to false analogies, and progressively becoming new rules and new norms.

If derailings of this kind are produced in the course of typical development and not only the course of typical behaviour, it becomes clear that the effects will be more profound. There are doubtless many reasons to guard against the old idea, at once anti-Lamarkian and anti-Darwinian, expressed in Cournot's well-known phrase, that the evolution of the species is 'a fecund teratology'. This idea has in essence been reprised under the guise of mutationism. But there is some advantage to distinguishing it from its interpretation by dogmatic geneticists. Experience presents us with many such derailings, most of which, naturally, have no chance of leading to the formation of a new species, but some of which can succeed, just as erroneous Latin led to French.

In some cases, the graft of an auditory vesicle or of nasal tissue will induce the formation of the bud of a member (Filatoff). When exposed to toxicity, the formation of well-differentiated eyes on the blastoderm, far from the embryonic axis, is observed. In regeneration, derailments, false sequences, are quite frequent, due, as Guyénot has noted, to the fracturing of different 'territories of regeneration', which then adjust as well as they can and only half-regulate, due again to stereotypical sequences. If, for example, the bones of an amphibian's limb are removed, they will not regenerate, and the gap will only be filled with connective tissue (Schaxel). If the boneless limb is then amputated, the stump will regenerate the complete limb and provide it with bone.

Of course, these kinds of experimental derailings are too crude to lead to durable formations. They are not spontaneously produced through small 'errors' of synchronism, or by small accidental disjunctions in sequencing. The 'release' of an inductive substance can be retarded, a buffer can be interposed between the stimulus and the competent tissue, and an inductor can act for slightly longer than it would normally do, thus producing various disproportions which change the form. Development, throughout all of the sequences that it implies, resembles a scavenger hunt or a complicated 'package tour': even the smallest hiccup demands an organisational reshuffling.

Variations due to Markovian accidents are most frequent in the morphogenesis of decorative organs. The designs on butterfly wings are, at least when compared with those of neighbouring species, precisely analogous to a tapestry produced by a Calliope-style machine in which chance is systematically introduced into the assembly. In every domain, aesthetic forms have a non-rigid unity, 'couples the precise with imprecision'.[24] They are midway between a meaningful language and the language of dreams where themes, while disjointed, remain just as expressive. In the case of human beings, it is the schizophrenic's associations, which tend towards being 'Markovian', that easily take on an aesthetic hue. Every ornamental form can therefore be expected to change. And in effect, the designs on a butterfly's wings can often be modified simply as a result of being exposed to an abnormal temperature during the pupal period. Thus exposed, the swallowtail butterfly becomes almost indiscernable from the Turkish *Papilio sphyrus* or *Papilio centralis*. In the same way, the plumage of certain pigeons closely resembles that of tropical species if it is exposed to a humid atmosphere immediately upon hatching.

> Goldschmidt has undertaken a systematic study of all of the known facts concerning phenocopy, in which, due to the action of external physico-chemical factors, a morphological character produced by a genetic mutation may be imitated.[25] The morphological effect of many of the known mutations of *Drosophilia* (the fruit fly) can thus be reproduced by coldness, heat, humidity or the presence or absence of a vitamin. Goldschmidt quite rightly concludes that these facts allow us to understand the habitual mode of action of genes. Temperature, by all appearances, acts on patterns* by influencing the rate of reaction, accelerating or retarding one formation relative to another. If temperature can imitate the action of a gene so precisely, it is because the gene also acts in the same way, as a catalyst, disrupting the temporality of the synchronisations or sequencing of formative rhythms.[26] The work of Goldschmidt constitutes major progress in biology, and a renovation of dogmatic genetics. Genes, like external physico-chemical actions, introduce an element of Markovian chance into the complex sequences of formative actions. By troubling or prioritising a direction at a critical moment, they reorient a whole chain in the same way that a simple slowing down or viscosity of intellectual acts suffices to explain certain characteristics of aphasia. But they certainly do not explain the morphogenetic themes themselves any more than the modes of drawing explain the contents of the boxes from which the draw takes place. In almost all of the most closely studied cases, their action is negative, such as those which produce notched, stunted, shortened or vestigial wings in *Drosophilia*. Their action is only ever indirectly positive, by effecting the prolongation of a positive action of a different kind, in the same way that, by surreptitiously setting the time on the clock back and thereby making the labourer work for longer, I become responsible for what they produce. This is why the success of phenocopies always directly depends on the moment in which, and the point at which, the disrupting agents intervene. The instant at which two or more themes are sequenced, and the decisive points of transition, must be grasped. Detailed research of the designs on the butterfly *Ephestia*'s wings by Kühn and Henke have revealed the moments and points at which the 'fabrication' of one of the themes of the pattern* comes into play for and is combined with others – the moments in which, therefore, a thread can be reoriented or removed from the whole fabric by affecting the loom. The pigmentation of flowers involves an analogous Markovian play of chemicals, gives rise to the same effects and allows for the same conclusions.

Setting aside these superficial ornamental formations, in which the norm of a species can widely vary without great risk, evolutionary crises due to Markovian derailing and 'fecund teratologies' are rare. In fact, species are stable, generally fixed for millions of years in a rigorously sequenced system in which aleatory draws are excluded on pain of death. The monotone repetition of individual ontogeneses escapes history for extremely long periods. It is thus instinctive behaviour alone, perhaps, that is Markovian to a limited extent. Its hazards can be deadly for individuals, but the species survives. An animal can be 'derailed', and take its eggs and young to be something 'to eat' rather than 'to care for', and caterpillars in procession or warrior ants can exhaust themselves by walking in a circle.²⁷ But the species itself does not die and, in general, no longer changes. It is rare for a displacement of instinct to engender a new specific formation. It is even more rare again for a displacement of instincts or formative activities at the most fundamental level, the level of ontogenesis, to succeed in provoking a viable general reorganisation and a new system of normality. Teratology is almost always mortal and non-fecund. All transgression is rigorously punished. Specific morphogenesis must escape history, and we must not be misled by the historical perspective that, in a single glance, takes in the rare evolutionary crises placed end to end. Even though 'fecund teratologies', displacements of normality and legality, are the rule in human society, and periods of long stability the exception, the opposite is the case in the 'societies' of morphological themes that are organic species.

Chapter 9

'Crossword' Formations

During normal periods of stability, a species is no longer 'Markovian'. It escapes history. Every species has a primary stability which arises from the fact that thematic potential is a-temporal by nature. But it also possesses a secondary stability, whose positive aspect is composed of a double negation – of an effort to eliminate open chains, and to close up immutable and, as such, repeatable cycles onto themselves. Species succeed in doing this in two ways: on the one hand, by interlacing and networking open chains; on the other, by giving regulative direction to these chains.

In normal morphogenesis, as we have seen, the task of formation is rapidly distributed to many presumptive areas, then to *primordia*, which differentiate themselves in turn, before encountering, adjusting to and collaborating with each other. Let's leave aside the adjustments and collaborations for the moment in order to consider the encounters. In this way we can compare the organism in formation to a sort of crystalline network, semantic rather than chemical, like a 'crossword' grid in which the words of the solution have more than one meaning and can be read both left to right and top to bottom.

INTERLACED EVOCATIONS

It is the totality of these encounters and interlacings that lead us to compare development to clockwork, to mechanised weaving or to a physical crystal, and which leads us to speak of the 'mechanics of development'. But the 'crystal', the 'mechanism' and the 'weaving' concern signifying themes; they are 'semantic'. Though the comparison with the 'crossword' puzzle – a kind of semantic crystallisation – appears quite fanciful indeed, it is very much closer to the facts, something that is easy to understand if the thematic and mnemic character of development is kept in mind. An area that is 'determined', but only in an abstract manner lacking a precise differentiating direction, is like the definition of one of the 'answers'. It must be confirmed by cross-checking, by its conjunction with other areas of determination. The ear or the eye is thus formed, not, certainly, piece by piece or by the slotting in of cogs, but by the convergence of themes. The facts of 'double assurance' are, whatever some embryologists say, the rule rather than the exception, at least in the sense that all local differentiation depends on a plurality of intersecting stimuli, on an evocative constellation.

In the same way that the solution of a crossword puzzle begins with one word or an easy zone that ripples or sometimes fans out – in a 'step-by-step' which is not, incidentally, strictly spatial since the words 'induced' by a certain answer belong to a single intuition – there are directive zones in the embryo in which primary inductors are at work: the blastopore lip, then the medullary groove, etc., and the directed zones in which the primordia are first constituted piece by piece or according to secondary themes. It is thus the case, for example, that the primordia which will become the circulatory apparatus appear in numerous locations and are then fused together to form the vascular network.

It is thus the case, then, that unrefined materials and states of a future form often precede, in behaviour as in development, the definitive form, which is born from these materials through a sort of transfiguring animation, just as in a crossword puzzle _G_MEM_O_ all of a sudden becomes AGAMEMNON. A. Gesell has given many examples of this borrowed from the psychophysiology of the infant. Motor behaviour is progressively established on the basis of the simpler motor themes which constitute its dynamic materials: 'A behavioural trait is first *nascent*, that is, simple and discrete; it becomes *assimilating* when two or more traits are combined; *integrative* when a trait acquires a second inseparable component; *coordinative* when a complex type emerges; and finally *synergetic* when all of its essential components have become interrelated'.[1]

From beneath the mechanist metaphors of biologists, which compare the organism to a liquid crystal, a group of physical fields or chemical systems, the better metaphor of the 'semantic network' or 'crossword' puzzle begins to emerge: 'Life', D. L. Watson writes, 'is a process or a method, an integrally connected series of operations, a contagious principle for the utilisation of free energy in reserve . . . The organising agent is not centralised, as philosophers like Rignano or Driesch imagined, but is diffused throughout the organism'. And the same metaphor holds for the most sophisticated of intellectual creations: 'Mental structures are as real and somatic as crystalline or fibrous structures . . . They are naturally produced by forces of induction, just as a crystalline embryo is created in the neighbourhood of an optical cupule in a morphogenetic field'.[2]

Obviously, no metaphor should be pushed too far. There is no grid, no set of white-and-black squares in the embryo – or rather, the fertilised egg itself begins by assembling, on the basis of its own primitive structuration, the

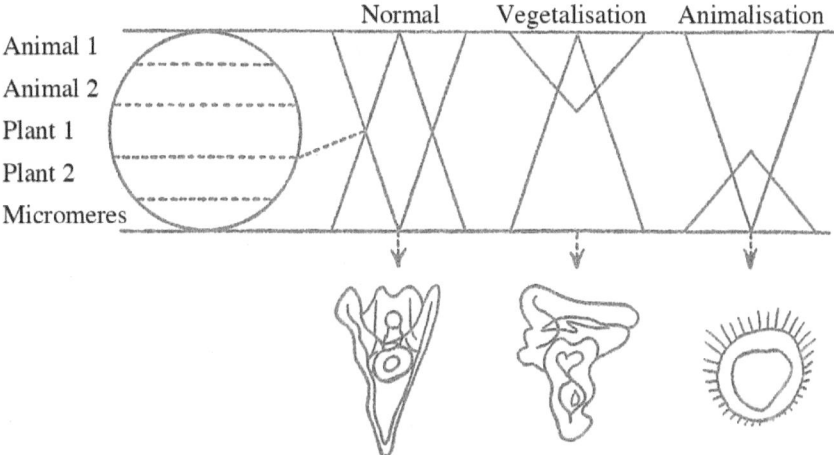

Fig. 9.1

equivalent of the grid or black squares, by establishing a multiple system of many entries, polarities, gradients and perhaps crystalline, physical networks, which then serve as the framework for semantic networks, for morphological discoveries and evocations. It is precisely this framework that allows the experimenter to act with such ease – allows the embryologists of the Swedish school, for example, who have modified in every way the double gradient of the sea-urchin egg, to disrupt the morphological filling-in of this 'grid' in a systematic fashion (cf. figure 9.1).

SEMANTIC NETWORKS AND THE PREFORMATIONIST ILLUSION

The value of the comparison lies in the fact that it provides a way of understanding biologists' persistent preformationist illusion while at the same time clearly showing it to be an illusion. When an easy crossword puzzle is solved without having to modify the grid, when the activity of filling-in seems to operate without any effort, as if automatically; when the right solution is attained quickly, as if the words were already visible in the fabrication of the problem, there is the barely noticeable feeling of an invention, an epigenesis, a creator's luck. It is nevertheless no question that the person who resolves such a puzzle, however easy it might have been, cannot be content with the claim that he simply engaged in acts of mental 'functioning'; by hypothesis, he did not know the solution, and, consequently, he must have engaged in 'noegenetic' acts, to use Spearman's terminology, analogous to those which a subject engages in an IQ test. The most straightforward 'eductions', such as 'Looking is to seeing what listening is to . . . (4)', which literally appear to leap into the mind, nevertheless require the theme to have been grasped and for the guidance of mnemic evocation. As Köhler remarks, many of the chimpanzee's 'problems' do not appear as problems to human beings. To happen upon a foreseeable and foreseen result in an IQ test, a crossword puzzle or a morphogenesis is in no way incompatible with an authentic morphogenesis. That all of the threads of morphogenesis intersect at just the right moment must not lead us to conclude that the first lineaments, the first primordia, the first axes of symmetry and the first polarities which are established in the fertilised egg are the equivalent of the starting up of a loom or the movements of a clock. They represent authentic 'eductions', and, like the first words discovered in the puzzle, they serve in turn to pose other problems of morphogenetic eduction. Thus, for example, the test cited above can be continued by asking, '(4) is to the ear what seeing is to . . . (5)' and then, '(5) is to a camera what (6) is to (7)'.

The play of intersecting eductions could continue in this way up until the point that completely empty formulations would have to be filled in, which would have already been the case in the preceding example if 'camera' had been previously uncovered by eduction from another chain.

SEMANTIC NETWORKS AND PROVIDENTIALISED HISTORY

The comparison with a crossword puzzle is particularly good, finally, for understanding the difference between morphogenesis and 'Markovian' history. As an inventive and creative activity, a normal morphogenesis in the egg or in the genes must not be taken to be preformed in the mechanist sense of the word, nor taken as 'providentialised', assuming the use of this barbarous word is allowed. The discoveries along each chain are expected, as are points of encounter and, consequently, future chains which will in turn begin from these points which are not yet actual. While the dividing egg is still a simple blastula, it would be pointless to search in space – and for good reason – for the point of encounter between a non-existent optical cavity and the ectodermic groove which will become the lens. *The eye is just as 'non-existent' in the blastula as the word 'eye' is in the still-empty grid of the crossword puzzle* or in the pages of the test not yet taken. But 'just as non-existent as the automobile or the CGT[3] are in the Paris of the seventeenth century' cannot be added. 'Providence', in the puzzle or the test, is in the mind of the person who created them, who knows the right response and who has assembled the correct intersections. In morphogenesis, providence is the sum of specific potential, semantic, non-material trans-spatial memory, the coordinated whole of non-actual themes.

> In political or technical history, there is no providence of this kind except in the reveries of theologians and metaphysicians, of Bossuet or Leibniz. For Bossuet, history is equivalent, in sum, to a morphogenesis: the Roman Empire prepared the world for Christianity in the way that gastrulation prepares for neurulation. In the same way, for Leibniz, Tarquin's crime is the equivalent of the introduction of sperm as the condition of development – the condition for the Roman Empire, Christianity and for the best of all possible worlds.[4]
>
> It is obvious, in fact, that human history is a great deal more 'Markovian'. It is not absolutely at the mercy of chance. Human efforts to resolve actual problems, reasonable in principle but dispersed, are submitted as a whole to possibility, to 'the nature of things' which vaguely appears to be a kind of providential norm and that traditionalists or progressive mystics imagine to be 'the correct answer'. But possibilities are not self-enclosed, interwoven, interlaced into a network the way that the semantic virtualities of a crossword puzzle or the morphological potentialities of an embryo are. And, aside from certain fleeting illusions, they are not themselves proven through cross-checking. In the morphogenetic efforts of history, there is no mnemic norm, no potential specified by subsequent realisations. History must invent itself according to a semi-fortuitous sequence of suggestions arising from various situations. Existentialism, the inverse of the providentialist philosophy of history, unfairly applies what is true about specific morphogenesis to human evolution and also unfairly applies what is true of human history to morphogenesis or specific behaviour. It denies the type, the ideal norm, the mnemic pre-solution and the directive role of natural instincts in behaviour and even – although this thesis remains implicit – in the formation of the human individual. Each stage – instead of being a set of already solved questions in a crossword puzzle, already conforming to a norm which envelops it – is only a new point of departure, in itself random, for pure freedom. Or rather, this is what existentialism would assert if it were a coherent doctrine, if it had not side-stepped biological problems.

In sum, with respect to development as much as behaviour, biological memory reconciles invention and preformation, providentialism and freedom. Mnemic intersections, oriented towards adjustment and reciprocal support, normally provide a durable stability to diverse species that is only broken under exceptional circumstances when a local deviation from a norm succeeds in producing another coherent whole, another 'semantic network', what the geneticists of the Goldschmidt school interpret as the general reorganisation of genetic equipment. It does in fact happen that someone working on a crossword puzzle can find themselves torn between an anticipated solution and another equally coherent and plausible successful solution. The evolution of a species results from this kind of deviation in which the reinvention that there is in all memory becomes pure invention. And it is for this reason that, if the normal morphogenesis of the individual according to a specific type does not resemble human history, the evolution of the species resembles a capricious history, reigned over not by a mnemic potential but by possibility *tout court*.

ACTIVE PRECAUTIONS AGAINST DERAILMENT

Let's turn now to the second procedure, quite different from the first, by which a species in a period of stability escapes from Markovian displacements. This procedure bears in a regulative direction. This regulative direction, superimposed on multiple chains, must not be confused with the primary action of potential which distributes the themes of development and initiates the chains – even though in reality the two actions are sometimes difficult to distinguish, and even though the regulative direction can in fact be considered to be a 'surveillance' exercised by potential in its residual unity. This regulative direction appears most prominently in instinctive behaviour, which facilitates its comparison with the psychic regulation undertaken by what the Würzburg psychologists call a 'determining tendency' or 'task', which orients particular associations and prevents derailment at key points where a 'Markovian' effect would be of concern.

The female whose male mate is aggressive risks being taken as an enemy; the animal returning to its nest risks being attacked by its occupants as a dangerous intruder; the young risk being eaten by their parents. Instinct often seems to take special precautions in accentuating the distinctive character of stimulus at these delicate points: the female adopts a submissive attitude; returning to the nest, the nocturnal heron (*Nycticorax*) offers a 'greeting', engaging in an 'appeasement ceremony' that involves raising three white feathers of its head.[5] In many other cases, however, the animal must, in order to avoid derailment, be the equivalent of a subject of the Würzburg experiments in whose mind the subconscious task transforms the impression made by an inductor word to the point that the same (objective) stimulus is no longer (subjectively) the same. Together with the stimulus, the task forms a new thematic totality. Cichlid fish, which raise their young in their mouths, must not experience them as 'morsels to eat'. Along with the stimulus, a regulative direction must form what Selz calls a 'total task'. This is, incidentally, what normally differentiates correct and meaningful phrases enunciated by a healthy individual from a Markov jargon: the general theme, or the semantic task of the projected phrase, guides the semi-automatic play of completed themes which only remain to be associated according to their resonances, and which are in fact associated in this way as soon as the dynamic surveillance of the theme falters, as is the case in aphasia.

The regulative direction also appears in morphogenesis strictly speaking. Embryologists distinguish in general – despite the fact that they only poorly understand their own vocabulary – between pure and simple evocation, and evocation in a field of individuation.[6] Thus neurulation can be evoked by chemical substances in a competent ectoderm without a normal sketch of the

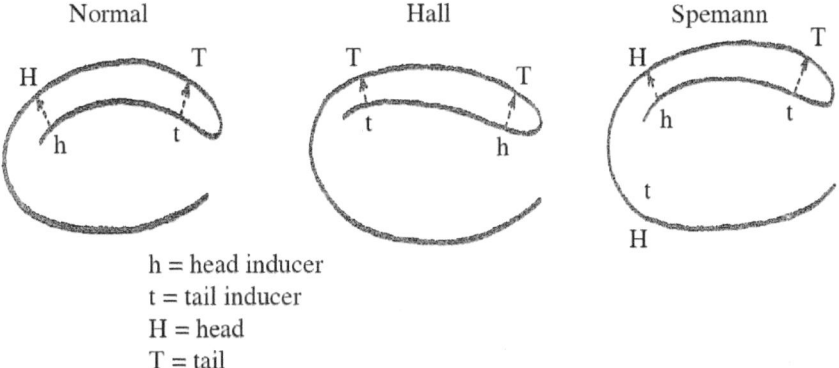

h = head inducer
t = tail inducer
H = head
T = tail

Fig. 9.2

nervous system, in which the 'head' side is distinguished from the 'tail' side, accompanying it. In a normal field of individuation, every tissue tends to be regionally differentiated not only according to the inductor-stimulus with which it is in immediate contact but also according to the typical location of the primordia which are normally part of the whole. Or again, when the experimenter systematically uses 'false' inductors' in an already constituted field of individuation, for example, in the grafting of – living or dead – tissues, which are normally inductors of the head, to the other side of the host which, normally, is the producer of the tail or conversely, the results are variable. Sometimes the inductor predominates, while at others the host's field of individuation predominates and directs the action of the inductor, as Spemann's experiment, in which a secondary inductor of the tail, grafted onto a secondary in the neighbourhood of the head of the principle axis, induced a head. Sometimes, finally, the correction is incomplete, as in E. K. Hall's experiment in which the inverted head and tail inductors only succeeded in forming a monstrous embryo with a double tail.

MARKOVIAN JARGON AND *TERATOMAE*

Teratomae, chaotic mixtures of tissues and even organs – hair, teeth, nervous tissue, bone, cartilage, amygdala – often found in ovarian cysts, the veritable Markovian jargon of specific morphology, appear to be the result of the action of evocative substances, beyond the control of any field of individuation, on adult tissues that have conserved various degrees of competence. These chaoses of organs and tissues never take the form of an individualised whole, and they do not resemble the monstrous embryos with which they are often identified.[7] Local, step-by-step correlations between tissues appear, however: smooth muscle associated with glandular cavities, cartilaginous islands with nervous tissues, as if secondary or tertiary inductors were at work, composing a purely regional order according to the effects of what Holtfreter calls 'vicinity' or tissulary affinity and in a fashion that completely conforms to the schema of a Markov chain.

Teratomae are generally benign, though some have a malignancy which makes them the relatives of cancerous cellular growths. It is natural to interpret the latter as formations which also escape from regulative dominance. And, in truth, this is not an interpretation but a simple description of facts, which teaches us nothing more than what is already known – they are cellular developments, normally dominated by the morphological theme of an organ or a tissue as these themselves are dominated by the field of individuation, which have been emancipated. It is even likely that cancerous proliferations imply an emancipation in the cell itself, at an even more elementary level, outside of the field of cellular individuation, of macromolecular particles, which we do not know whether to call 'viruses' or 'constituent proteins' at odds with their milieu.

Cancerous formations are thus no longer Markov jargon at a relatively elevated level of organisation, in words or completed expressions, like benign tumours; they are a Markov jargon at a lower degree of organisation, where the cellular themes themselves are disjoint even though the malign cells remain, all the same, cells and even maintain the most frequent characteristics of the tissue from which they arise.

Every stimulus and every inductor, in itself normal, can be a factor of disorder if it acts and evokes a competence in the absence of a subordinated regulative direction. It is a well-known fact that analogous or identical substances can be at once feminising hormones (oestrogen), agents of induction and carcinogens, thus acting in both normal and pathological morphogenesis. In all likelihood, 'false' proteins, like false stimuli or fake keys, can participate in cycles of cellular formation. The modes of action for 'true' and 'false' inductors are often completely indiscernable. L. Loeb, Needham and Waddington have pointed out, for example, the analogy between the transmission of cancers by injection of cellular fluid – which leads to the attribution of a filtering virus to them – and 'homoio-genetic' induction, by which the secondary neural plaque, once induced by substances emanating from the primary organiser, becomes capable of inducing a secondary neural plaque and this in turn, the 'fluid' of neural plaque being able to *propagate* neurulation just as well as an implanted graft.

The inevitable use here of the words 'normal' or 'pathological', 'good' or 'bad', 'true' or 'false' is indicative, and it proves precisely that normal sequences are dominated by a regulative direction which prevents them, except by accident, from being Markovian, and which makes them resemble meaningful phrases. A Markov chain, an association of ideas, an open sequence is neither true nor false. The true and the false – and, in the Würzburg experiments, the *feelings* of the true and the false in the response given to the inductor word, the feeling of a possible judgement of response – only appears if a task is imposed and dominates associations. Whether it succeeds or not, it is this presence alone which justifies the use of the words 'true' or 'false', 'good' or 'bad'. A signal stimulus can already be true or false to the extent that it serves a sequence whose division is substituted for a dominant theme and which only has the value of a substitution.

When a dog, conditioned by a stimulus, salivates even though the experimenter has not confirmed the stimulus; when a fish lets itself be misled by an artificial releaser* and reacts to a lure; and finally, when an organic tissue responds to a hormone or a protein which only resembles the hormone or the habitual protein and the subsequent induction is pathological, the words 'false', 'lure' and 'abnormal' only have a meaning because, above the sequences, there is a theme or a dominant task which sometimes succeeds in redressing and normalising the sequence – the dog stops salivating through internal inhibition, the field of organic individuation normalises a regenerating agent that was at first atypical – and sometimes fails, leading to death, sickness or monstrosity, except in those rare cases in which error is the origin of a viable evolution and a new specific norm.

Chapter 10

'Spectacle-Spectator' Complexes

Biologists apply the term 'allesthetic', that is, 'being perceived by others', to characteristics which are only conceivable in an individual, in their origins and their biological role, through the intermediary action of the distance receptors – the eyes, ears and olfactory organs – of another individual. For most organs and organisms, morphogenesis, more or less adaptive, explains or rather is explained by the inherent necessity of their own form or of their form in relation to its milieu. But for a large number of organisms and organic features, morphogenesis can only be explained because the *appearance* of the organism in the perception of other organisms was an essential factor in its evolution. Let's consider some examples. With the exception of the primates, mammals display no colour other than black and white and intermediary shades of grey, russet and tan. True reds, purples and blues only reappear in primates, monkeys and humans either in the organism itself (in the mandrill, for instance) or in their extra-organic productions. And this is certainly not by chance if the primates are also the only known mammals to possess a strongly developed sense of colour. The vibrant colours in many birds, lizards, fish and entomophilic flowers also imply the existence of sensations of colours in organisms concerned with 'appearances', and for whom they function as signal stimuli. The spectacle-receptor correlation extends into the details. Bees, which are blind to red, nevertheless perceive ultra-violet, thus correspond to the colours of the flowers that they pollinate. Red, on the contrary, is a good stimulus for diurnal birds; it also often appears in fruits the seeds of which they scatter, or in the flowers which they pollinate.

The animals from the depths of the ocean, which possess the most refined luminous organs also possess the best developed eyes, and it is also remarkable that the structure of luminescent organs, with their photogenic substances, reflectors and lenses, correspond to the structure of the organs sensitive to light whose functioning is their complement. Similar correlations between sonorous organs and sonorous receptors, or between odiferous organs and their olfactory receptors, are found in insects, birds and mammals.

These correspondences are closely analogous to those which exist between male and female organs, which exclude the possibility of being explained by a single moment of natural selection. Selection seems to be able to explain the correspondences between entomophilic flowers, but it is hard to see how it would explain the correspondence between photo-emitters and photo-receptors, and impossible to see how it would explain the similarity of male and female organs. To be more precise, selection – if there is selection – must have to act on the whole, on the 'spectacle-spectator' complex, as it does on the 'male-female' complex, whose correspondences therefore remain to be explained on their own terms.

What is possible in the case of flowers and insects, namely, that one of the already existing terms presided over the selective formation of the other, is clearly excluded when it is a question, not of semi-historical adaptation – whether parasitic or symbiotic – of two very different species, but a 'stimulus-reception' adaptation *internal* to a species. The spectacle-spectator complex represents a kind of double and conjugated feedback* circuit in which each of the two terms functions to regulate the other. As in all cycles of functioning, and all the more imperatively, the deployment of terms must be perfectly conjoined; one cannot be the pure cause and the other pure effect. Dalcq has observed that the patterns of pigment distribution external to the vertebrae are the result of cells which emigrate from the neural crest during embryogenesis.[1] The same process could thus give rise to both perceived form and the apparatus that supports perception at the same time.

ORGANS 'FOR BEING SEEN' AND BERKELEY'S PRINCIPLE

Allesthetic characteristics seem to verify Berkeley's celebrated principle 'To be is to be perceived'.[2] As J. Huxley has pointed out, light and colour in the strict sense, as a psycho-biological experience and not 'photonic radiation', 'did not come into existence before there were animals with eyes'.[3] In reality, the partial success of the formula proves that it does not possess a general value. The great majority of morphological characteristics do not depend on perception by other beings. Organs 'for being seen', to use A. Portmann's expression, are quite numerous but exceptional all the same. The viscera and even the exterior forms of a crowd of organisms must be directly understood. They depend, perhaps, on a sort of auto-vision, if this very inaccurate expression can nonetheless be used to designate the characteristic of the 'absolute domain' of the organism; it certainly does not depend on the vision of another.

It is clear, incidentally, that the morphogenesis of organs 'for being seen' themselves must operate on the basis of forms which do not possess this characteristic. It is another layer of clothing, formal attire, and sometimes a disguise. Perception and the emission and organisation of 'perceptibles' represent an elaboration of organisms, one that is certainly effective but secondary. Perceptual consciousness is an event, a new fact in the story of life, but it cannot be a primordial fact. Image-forms, forms-for-another, can only be a particular case of forms more generally.

ARTIFICIAL AND NATURAL SCENERY

Let's consider more closely what constitutes a 'spectacle-spectator' situation and, correlatively, what perception itself is. Since the theatre is a typical example of a 'spectacle', let's say that A, B, C, D, E and so on represent elements of the scenery perceived by the eye of the spectator.

We have the strong impression that it is the eye, the gaze 'emanating' from the eye of the spectator, which creates the unity of the scenery, formed from the distinct panels A, B, C, D. And broadly speaking, this is indeed the situation. The set designer sets the scene by repeatedly adopting the spectator's point of view, eventually establishing the place that will be occupied by the future spectator, the place from which he will be able to 'cast his gaze' across the whole scene. It is the limit of this unity brought about by the gaze that it no longer subsists, so we think, in the real object, except as a piece-by-piece coherence without any true formal unity. Each element of the scenery is just a cumbersome panel that needs a special truck in order to be moved around.

Without going as far as Berkeley – who was a philosopher, if not a man with any common sense – we can at least go as far as believing that the form of the

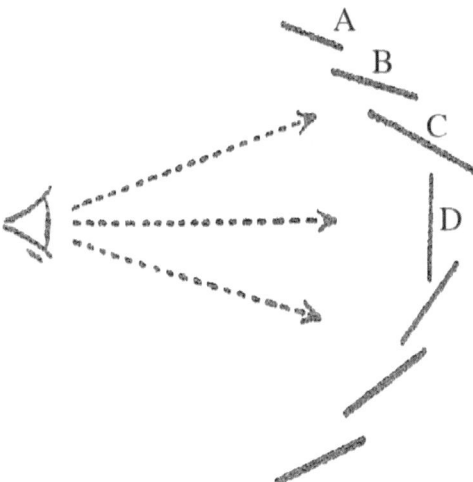

Fig. 10.1

object is no longer seen as any more than an indirect, operative reality subordinated to a physical coherence which is exercised by one molecule on another.

Now let's suppose that, instead of representing scenery in a theatre and a spectator in the audience, the schema represents a peacock which parades its feathers in front of its peahen, or any animal or organism whatsoever presenting its allesthetic features to another. It now becomes apparent that a third term must be admitted, an intermediary way of being between the image-form and the pseudo-form possessing a piece-by-piece coherence. Organic form in itself, independent of any gaze directed towards it, possesses an authentic formal unity. The peacock's tail has as much right to be considered as a formal and thematic unity in itself as the bird's unseen viscera do. Morphogenesis or the organism's behaviour, even if it is conceived to have no allesthetic features, possesses the character of true form, goes entirely without an image-form. Unicellular organisms, young embryos and plants do not have eyes, and no eye can see them; they are, nevertheless, active unities, profoundly different from the entirely conventional false unity of the scenery in an empty theatre.

PERCEPTION AND TRUE FORM

Incidentally, if perception itself is considered, rather than the spectacle-spectator *situation* and the mise en scène of an organic parade or a theatrical representation, the completely false character of the preceding schema is quickly realised. As we have known since the ancient Epicureans, the eye obviously never *casts* a gaze or *sweeps across* external objects in order to unify them into an image; it does not resemble a television's 'electron guns'.[4] The retina or visual cerebral area on the contrary *receives* waves and photons. Above all, the retina or visual area does not resemble a material screen; like all living tissue, it is already by itself a formal unity which belongs to the organism, it is an 'absolute surface' and not a group of physical objects arranged on a surface. This is necessarily the case in order for the physical pattern* of

Fig. 10.2

luminous waves to become a conscious image rather than remaining a pure physical structuration.

It is clearly contradictory to still want to explain visual sensation in terms of scanning* or cerebral sweeping, on the basis of a renovated schema of the mise en scène of perception, transported into the skull. There is therefore no reason to stop here. The set $\{a, b, c\}$ exists absolutely in a visual area as a formal unity without any need for a new scanning* to be grasped in itself. This 'to exist-together-absolutely' is given in visual sensation by the living tissue which is thus primitively defined. It is therefore absurd to explain existence on the basis of perception, and the conscious image, which is only explicable in terms of the mode of existence – as absolute and primary form – of the organism.

The perceptual image, along with all features 'to be seen', presupposes a living form and primary organic characteristics. It is the whole organism that is capable of 'perceiving', that is, of becoming conscious of itself, no matter what composes the ensemble of external stimuli, because the whole organism is a surface or an absolute volume, a form which exists by itself and which only has to present itself to this *ensemble* in order to participate in its mode of being a true form. In the same way that any white surface, tablecloth, sheet, serviette or wall can function as a screen for the projection of a photograph

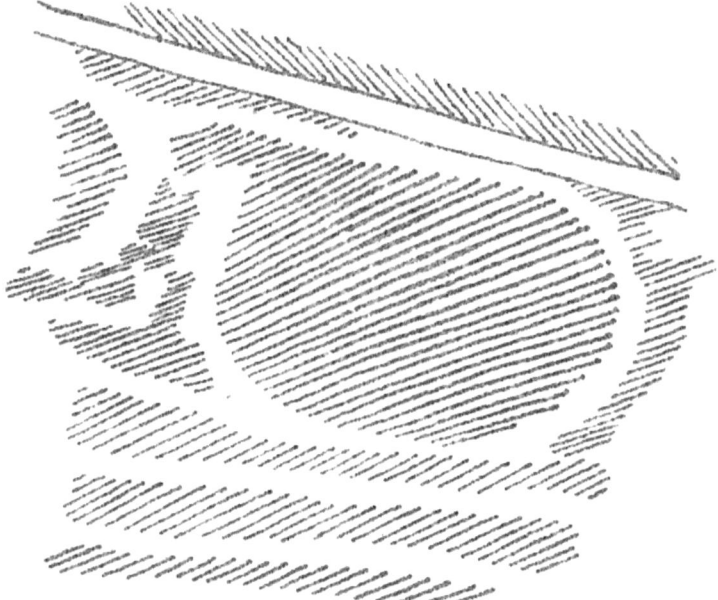

Fig. 10.3

or a film, any organic tissue can in principle – at the cost of a secondary technical adaptation, such as those which induce the retina or the inner ear and the corresponding cerebral centres to become organs turned to the reception of sonorous or luminous waves – become capable of perception, capable of being modelled on external stimuli by transforming them into a conscious image. And if all organic tissue is capable of perceiving, it is naturally also capable of inducing in itself a spectacle of being seen. Containing in itself the equivalent of an eye, it can be offered to the eye of another organism.

The mystery of 'allesthetic' forms is dissipated to the degree that this point is grasped. Let's consider a precise example of what passes for one of the enigmas of nature. The feathers of the Great Argus pheasant are used in mating dances. They are without question 'organs-to-be-seen'. During the dance, the male's wings are ranged in front of it, like a large bell oriented towards the female, while the rest of the body remains hidden. The barbules of each plume compose a complex but regular design composed of bands, points and spots, which give the impression of solid globes due to their light and dark shadings. Now, this design has arisen through a single variation of pigmentation along each barbule.[5] If you the reader try to draw a circle directly, with reference to nothing but parallel lines – and we ask that you experience this first-hand – you will come to perceive the extreme difficulty of the operation. And the difficulty would be even greater again if you were to try to give the impression of relief through the use of shading. The organism cannot be working blindly as it forms this striking design. The plume is like a formative retina, analogous to the receptive retina, in absolute possession of its own form and thus capable of directing formation without being obliged to take a step back, like the painter of the theatre scenery, and place itself from time to time in the position of the future spectator in order to judge the effect. The theme of the design must exist absolutely and dominate the chemical phenomena which realise it through each barbule in the same way that the retinal image in perception exists absolutely and dominates the chemical phenomena which it produces in each of the retina's cells. The formation of a design on the male's plume and the perception of the design by the female are at root the same phenomena of 'absolute surface', the one being produced on the plume and the other on the retina or its cortical correspondent. This is proof that the plume, or any living tissue whatsoever, is not essentially different from a retina or a cortex, the proof that every organism in development is a field of formal themes which realise themselves, which shape it directly, that it is modulated by formal themes, sensory areas simply possessing the particular property of being modulated by already realised structure in the external world by other organisms or other natural complexes.

DIRECT FORMATION AND FABRICATIONS
BY THE EYE AND THE HAND

In the problem of organs 'to be seen', and in the problem of perception, we rediscover exactly the same situation that we have defined with respect to behaviour in its opposition to functioning. It is naïve, we remarked, to believe that human locomotion could be explained *first* as neuromuscular functioning, and *then* to marvel that a protozoon can improvise pseudopods directly formed from its protoplasm for its locomotion, even though human locomotion, which involves putting one leg in front of the other, is itself nothing other than the manifestation of the primary behaviour of a cellular network in the nervous centres, a network analogous to an amoeba or colony of amoeba.

Nervous tissue is, after all, nothing other than an organic tissue like all the others and comes from a part of the embryo which would have been able to produce the banal epidermis. In the same way that every cell and tissue is capable of behaviour, every cell and tissue is capable of perception. The form of a protozoon, as much as the form of a human embryo, or a human adult to the extent that it has not entirely become a functioning machine, comports *itself* and sees *itself*.[6] The development of the higher species is not the result of the revelation of the properties of behaviour, auto-conduction and auto-perception. These fundamental properties belong to everything which is an organism, or, if you prefer, all true form – including unicellular organisms and viruses. These developments only consist in the advent of behaviour and perception better suited to the external world, which is not only auto-conduction or auto-perception, but conduction and perception *in a given milieu*, and this is thanks to the process whereby one part of the organism becomes specialised as an organ for the reception of external stimulus and behaviour according to this stimulus, the rest of the body being relegated, even if never completely, to being machinery.

In human beings, the brain thus concentrates almost all of the organic possibilities of behaviour and perception. In human beings, spectacles, like tools, are assembled by the intermediary of the cerebral cells, and it is also through this intermediary that spectacles are seen, and tools are used. Eyes, legs and hands, as functional organs, are subordinated to the primary behaviour of neurons or the neural networks that direct them. Considered as an organism in action, a man giving himself a tattoo is not essentially different from an organism forming ornamental organs. The sole difference is that the theme of ornamentation first appears in a human being as an idea of ornamentation in the cerebral tissues before being carried out by the hands and under the control of the eyes, themselves controlled by other zones of cerebral tissue which thus concentrate the activity of ornamentation, as they do for almost all other behaviour.

But in plants and in animals (in particular the lower animals), and also in human embryos, this specialisation has not taken place. It is the organism as a whole which is given over to behavioural themes, or to utilitarian or decorative formal themes, which are played out in the tissues like shadows on the movie screen, or rather – since tissues are not passive receptors – like the marks and colours of a painting which is guiding itself, or controlling the painter. The organic painting, being not a physical surface but an absolute surface, contains in itself at once the equivalent of the eye and the hand.

The astonishment that greets decorative organs and 'organs to be seen' forming themselves without an eye or a hand is astonishment if not before a false mystery, then at least before a mystery artificially complicated by an arbitrary inversion of the facts. We cannot prevent ourselves from *starting* with the situation of the man who paints, with his hands and with the help of the judgement of his eyes, theatre scenery, that is to say, in those cases where the spectacle-spectator complex remains complicated by superimposed external circuits – as if the situation of a man painting scenery or helping his wife choose a dress at the couturier were simpler and more natural than the situation of the female before the mating dance of the Great Argus pheasant. In fact, it is clearly the human situation which is the most complicated and which must be explained on the basis of the situation of the pheasant, which itself must be explained by the situation of the protozoon. The couturier who makes or directs the manufacturing of the dress has to have conceived of a pattern first produced by the nervous cells, which themselves utilise the properties of the absolute surface of every living tissue. Manufacturing, which appears less mysterious than organic development, is still development, development through an external circuit, complicated by nervous and social relays, of organic and extra-organic techniques. It is curious how widely held the belief that manufacturing is less mysterious than organic formation is even though manufacturing presupposes an organism that formed the circuits of manufacturing, drawing on the morphogenetic properties of their nervous system to do so.

The illusion arises because of the fact that the eye, the hand, the brain, visual consciousness and motor consciousness *are given* and that such an act of manufacturing is straightforward for the artisan, or for the daydreamer who imagines themselves in the artisan's place. But the *total phenomenon* of manufacturing is clearly more complicated than organic development. If organic tissue and its character as absolute surface is taken to be given in the same way – we mean, given in itself, lived for itself – development or organic creation is at least a degree less clear than technical manufacturing. It is the property of auto-conduction of organic form which explains manufacturing and which is the foundation of every technique, just as it is the property of locomotive behaviour by protoplasmic deformation, present in unicellular organisms and their colonies, that is the foundation of the locomotion of the higher animals and even of the human techniques for transport. It is the property of absolute surface of every living being which is the foundation of perception by distinct visual organs, and even of the human techniques of the theatre or the cinema.

There will be no shortage of accusations that a philosophy which explains manufacturing on the basis of organic creation, ocular vision by the organic property of 'absolute surface', is explaining *obscurum per obscurius*.[7] The preference is to explain organic creation, morphogenesis and primary organic form by manufacturing without realising that manufacturing already presupposes an organic creation in the nervous tissues of the manufacturer. There is here a veritable intellectual aberration and, if we might add this, a naïveté entirely analogous to that of the child who says, 'I know the difference between boys and girls, but only when they're dressed'.

The passage continues, from the protoplasmic network to the cortical network with its cortical 'retina' and 'hand', and then from the cortical network to the eye, the hand and the tool and then from the eye, the hand and the tool to the machine and the factory. Technicist philosophies return to the eye and the hand but often neglect to return to the organic networks on which everything nevertheless relies.

The unified and oriented behaviour of the hand at the end of an adult's arm relies on the behaviour of the 'cortical hand' of the motor zone. In order to perceive a spectacle, the human eye would be useless without the occipital visual zone. In turn, the cortical 'hand' and 'retina' only represent particular cases of the property of auto-conduction and absolute surface of all living tissue. The cortical 'hand', which directs the movements of the hand organ closing around the tool's handle, manifests no other property than the property of the protoplasm of the amoeba, capable of closing around a bacterium which will serve as food, making use of and making converge the energetic properties of contractile proteins. The cortical retina, and its absolute auto-vision, only manifests, in its specialised form, a property present in the protoplasm of the cell and without which this protoplasm would possess no true consistency and would only be the sum of its chemical processes.

Successive complications of organic and extra-organic technique are certainly not without their own morphogenetic effects. While the human hand and eye, human tools, machines and spectacles may depend in the end on the 'amoebic' behaviours of cerebral cells, they are capable of attaining forms that amoebic behaviour, left to itself and without a secondary assembly of relays, would never be able to create. This is precisely why, as we have insisted, the advent of the eye capable of seeing forms and external colours was the correlate of the advent of 'spectacular' forms, in addition to utilitarian organic forms. In the same way, the advent of the human hand much later in evolution, and even more profoundly, of the whole human symbolic apparatus controlled by the brain, allowed, in the same way, for an enormous morphogenesis of techniques and extra-organic cultures, which have in their turn exercised a recurrent effect of great amplitude on the organism and the human consciousness that supports them. But it must be understood that the continuity between them is less than perfect and that everything depends on the character of the absolute surface of organic forms.

Chapter 11
Forms I, II and III

The situation of the biologist before an embryo is stranger yet again than that of the physicist trying to decipher the secrets of the physical world. Books written by physicists for the wider public often conclude with the author drawing in close to say, 'Despite the progress of science and the ingenuity of our researchers, the universe remains full of mystery. But the most remarkable of all mysteries is still that we, a part of this universe, are capable of interrogating nature and understanding it to a significant degree'. But what do embryologists say? Several years ago they were themselves in precisely the same situation as their very objects of study. When an embryologist studies a development, we find one organism in the presence of another, one that is simply at a more advanced developmental stage than the other that it observes. When an embryologist publishes some observations concerning a developmental phenomenon, following his name with a list of his publications in the academy and elsewhere, he forgets the most important situation in which he finds himself, namely that of an 'ex-embryo'. Is it therefore rational to refuse all consciousness to the embryo at the 'gastrula' or 'neurula' stage under the pretext that it only appears as an object to the organism – the embryologist – that observes it?

THE EMBRYO AND THE EMBRYOLOGIST

The embryologist – *in other words, the grown-up embryo* – certainly observes the young embryo under study in a way that the embryo does not observe the biologist on account of the fact that it does not yet have eyes and its brain is no more than a gutter. But this superiority of the biologist is, first of all, short-lived. Nothing prevents the observed embryo from becoming, in its turn, an eminent biologist or great neurosurgeon who will observe, with a profound sense of superiority, the now-deficient brain of its earlier observer. And above all, this very observation clearly takes place on the basis of the subjacent condition that a development has occurred in the organism of the observer, who has passed from the state of a fertilised egg or 'neurula', occupied some thirty years earlier, to that of am adult organism, endowed with eyes and a functional nervous system. The scientific consciousness of the biologist clearly takes place on the basis of the consciousness or formative subjectivity of the embryo that he once was. The form of his scientific schemas have as their condition the primary form of their own organism. It would be absurd if he refused this absolute 'for itself' that he himself, in himself, makes use of to outline and speak of the organism that he is studying.

There is nothing mythical in considering the observed structure of the embryo to be the spatial manifestation of a form truly 'for itself', in possession of itself. To refuse this would indeed be to condemn oneself to a mythical belief. The biologist who in fact persists in envisioning the embryo that he observes as a pure ensemble of phenomena deprived of auto-conduction and its own conscious 'subjectivity' is condemned to the belief that his own consciousness miraculously appeared through an 'animation' – but granted by which god, and at which moment? – which abruptly transformed a collection of chemical phenomena into a consciousness capable of meditating on the role of chemical phenomena.

Let's therefore accept the following facts without further stipulation: that the embryo knows embryology better and more directly than the embryologist, that the liver knows its role and is more competent in hepatology than Claude Bernard or Cannon, and that Harvey's heart knows how to ensure the circulation of blood before Harvey's brain is advised that blood is circulating. Equally, let's accept the fact that an atom knows atomic physics better than Bohr or de Broglie.

The word 'knowledge' [*savoir*] would only be a metaphor if it were claimed that it applied to either collective phenomena without internal unity or to machines which have been assembled according to the 'knowledge' of an engineer but which, in themselves, can only function. It would be pure metaphor to say of a set of scales that they *know* [*sait*] horizontality better than a painter who represents them. It would also be pure metaphor to imagine that this knowledge of the mode of a consciousness or a *second* form implies perception and schematisation, reflecting and representing objects exterior to itself. This 'knowledge' is primary, like the knowledge inherent in a perfectly mastered activity which unfolds with perfect precision, without any need for auxiliary images.

FORMS I, II AND III, AND ZERO-FORMS

It is useful to designate these two forms as Form I and Form II. But it must be understood that one just like the other is 'for-itself', or 'conscious', and that Form II is only a very particular case of Form I, one in which Form I is arranged into the organs of reception of an exterior pattern*, or into organs of control according to a motor schema acting on auxiliary relays. This particular arrangement leads, correlatively, to consciousness becoming perceptive or schematising consciousness, and the 'for-itself' of the form becoming 'consciousness of objects' while seeming to attribute its 'for-itself' to exterior beings which reflect back on it, in turn becoming 'conscious images', to the point that philosophers like Berkeley and the neo-realists are deceived and come to imagine that the entire being of these objects only possesses the status of an image.

Correlatively, Form II produces, exclusively in the human species, a third type of form, Form III, thanks to the techniques of language and symbolism. Form III appears when utilitarian perception which, in animals and human beings insofar as they lead an animal life, only serves as a signal or an index of instinctive life, changes its role, and when the signal becomes a symbol, able to be independently manipulated and to be detached from every context of vital or immediate utility. On the symbolic plane, this passage to Form III, as Cassirer and S. Langer have shown, takes place for humanity in every domain. From the pure signals that characterised their primitive reality, vocal gestures become words, allowing for the combination of inactual ideas, narration, the forging of fictions, explanation, analysis, speculation and play; emotional gestures, no longer simply playing the role of signals for one's companions, become ritual gestures with a timeless signification: the origin of religious acts and mythology. The expressivity of perceived forms is detached from the vital situation, allowing for artistic play and the free creation of aesthetic forms, and they no longer simply serve, as 'organs for being seen', the needs of sexuality or social stimulation, and which do not only express, as has often wrongly been believed, the current emotions of the artist, but live their own life and develop by themselves even though they are always naturally attached to possible perceptions, that is, to Form II, which are themselves always attached to Form I of the organism.[1]

It may be useful, finally, to designate all pseudo-forms as *zero-forms*, which are only masses, or equilibria, or step-by-step assemblages, and which only possess the reality of their constituents.

TYPES OF FORMS, TYPES OF CONSCIOUSNESSES AND TYPES OF SPACE

By considering the three types of true forms, derived from each other, the characteristic of all morphogenesis as a conquest of space and time, a conquest which is also a creation, can be much better understood. We have already seen that Form I of an atom or a molecule is not a structure but a structuring activity, an action 'to fill up space',[2] according to Bachelard's expression, or again, it could be said, since space does not pre-exist as a recipient, an action to constitute a domain of space. Giant molecules and viruses constitute greatly extended domains, but of the same kind. The organic techniques of membranes and ducts allow for a new conquest. They allow for the annexation, by Form I, of zero-forms, transformed, by canalisation and imposed sequences, into organic machines. Nervous systems, at once 'machines' and 'domains', allow, thanks to their modular networks, for variable integration, the emergence of Form II or 'perceived' forms, thanks to which extension seems arrayed before a subject which thus enters into the possession of an *Umwelt*. The 'subject' or 'perceiving consciousness' thus seems detached from extension and duration like a kind of point of view even though it is reality itself and the absolute of extended and temporal form. Consciousness seems opposed to form, when in fact the two words are synonymous and interchangeable, and consciousness is only, as form, constituted space and time.

THE SOLIDARITY OF FORMS I, II AND III

The three types of forms are indeed distinct. States of the universe have existed in which no Form III existed, and states of the universe have existed in which there was no Form II, in which no organism had yet developed sensory organs. In contrast, Forms I have always existed, at least since the universe has existed, since a form-zero is only a multiplicity of Forms I edge-to-edge.

It is indeed necessary to grasp that Form I is fundamental, and that forms II and III would be inconceivable if they were not based on Forms I, of which they are only particular cases. The three types are distinct, but each is united with the one preceding it. The solidarity of the three types of forms is apparent in a range of phenomena. Let's even put aside the obvious fact that the forms created by human culture, entirely in keeping with non-biological norms, presuppose the human organism and psychological consciousness. The whole of Form II manifests Form I, which supports it. It was probably psychologists who first recognised this. Psychological life is incomprehensible if, as the associationists of the eighteenth and nineteenth centuries claimed, it is separated from biological life, or if it is reduced to isolated sensations and images, filing across who knows what screen in front of who knows who, clumping together and drawing the attention of a recipient-consciousness. Psychic life is Form I as much as it is Form II. There is a thought without images, a knowledge without the images, tendencies and themes that guide the superficial play of Consciousness II. At root, the fundamental theses of the Würzburg school, the psychology of language difficulties inspired by Jackson's 'verticalism'[3] and psychoanalysis are in accord.

What psychologists call the dynamic unconscious is form, thematism, the primary consciousness of the organism manifesting itself in secondary consciousness. In terms of the theses of German romantic philosophy, which declared the unconscious older than consciousness and saw in it the primitive givenness on which consciousness was subsequently established, it is clear how easily they could be transposed into the much more appropriate language of primary and secondary form. Cyberneticians and mechanist neurologists, who stake their ambitions on the construction of 'mechanical models of consciousness', have not taken account of the fact that before constructing a mechanical model of consciousness, they would have to have made, if such a thing was possible, a 'mechanical model of the unconscious'.

For their part, biologists assert the same continuity in a thousand different ways. Let's only recall the innumerable cases in which morphology and physiology on the one hand, and behaviour related to perception on the other – in short, Behaviour I and Behaviour II, internal circuits and external perceptual circuits – are clearly in continuity, and the no less numerous cases in which they can substitute for each other, perceptual behaviour leading to the same result as directly formative behaviour. Sexuality, for example, involves anatomical formations, organic instincts and psychic instincts at once. And it is remarkable that hormonal themes intervene in both the genetic control or the shift of anatomical sex at an early stage and in the later stimulation of instincts in the adult, where it is related to psychic schemas of perception and behaviour. Both organs and tools controlled through perception substitute for each other.

David Lack has recently given a great example of this in his study of the finches of the Galapagos Islands (*Darwin's Finches*).

Darwin was already struck by these birds. They certainly derive from a continental species. Today, they form four distinct genuses and fourteen separate species which are insectivores, omnivores or feed on leaves and flower buds.[4] One species chases insects into cracks like the woodpecker. All of the species have modified the form of their beaks in keeping with their diet, but the 'woodpecker' type has developed a behavioural instinct in place of a morphological modification: it makes use of the spine of a cactus held in its beak to chase insects to the depths of the crack.[5]

The protection of the young of various species – leaving aside the human species – takes place, as if indifferently, by organic dispositions (whether those of the parents or the young), by instinctive constructions, by protective behaviour directed by perception, by instinctive protective demands or by some combination of the above.

As one might expect, it is above all in the domain of forms and coloration, whether mimetic or for courtship, in which the most beautiful examples of the solidarity of organic forms and behaviour oriented by perception are found. Take caterpillars, for example, which adopt a position corresponding to their colouration, stomach or back up, displaying as a result a pale colour, and which lie diagonally or longitudinally, following the direction of pine needles, displaying as a result diagonal or longitudinal bands in keeping with them.[6]

More generally, perceptual phenomena – because they are expressed by morphological phenomena – *can always be repurposed in the circuit of morphological phenomena*. Many fledglings – human babies included – do not flourish, or even survive, if they are not caressed. A bird's laying is regulated by the perception of the eggs already laid. The perceptions-stimuli of sexual instinct, whose threshold of effectiveness depends on the primary organic state, induce in turn a primary organic state. The eyes can sexualise just as well as hormones. Perception has a 'nutritive' as much as a 'signaletic' role.

> [The] ovarian follicles of the female starling, when isolated from the males, do not exceed a diameter of three millimetres. The presence of a male in a neighbouring cage makes it grow to five millimetres, and when the female is placed directly in contact with the male and close to the nest, it grows to ten millimetres. And this auxiliary action of psychism exists in numerous other species.[7]

STIMULI I AND STIMULI II

It is necessary here, not to correct, but to clarify what was said much earlier about signal stimuli. Causes of the classical kind, that press from edge to edge, are, as we have seen, only at work in an organism to the extent that the organism has been transformed into a pure machine. These 'causes' only concern zero-forms. As for signs, they concern human beings as creators of language and culture, that is, Forms III. The true organic 'efficacities' are signal stimuli. But the notion of signal only applies, in the strict sense, when there is perception, that is, to Forms II. And yet, because not everything could not be said at all at once, we have provisionally considered Forms I and Forms II together from this point of view, admitting that 'efficacities' for both were stimuli, and even signal stimuli, not causes. The word 'signal' is clearly something of a metaphor if actions on an organism or a tissue, rather than on a sensory organ, is at issue. For a signal to be a signal, it must be perceived. As Buytendijk emphasises, it is difficult to speak of a signal when, for example, the action of a vegetal hormone is at issue.

And yet, the complete solidarity of forms I and II, of primary and perceptual consciousness, in good part justifies this confusion. A stimulus or a hormonal signal is not a perceived signal, but the difference is not essential, as is shown by a great many intermediary phenomena, about which one no longer knows whether to speak of perception or more direct action. When the development of the gonads of a bird is unquestionably stimulated by the sight of a sexual partner, it is unquestionably a matter of perceived signal stimuli. When it is stimulated by light in general, it is already difficult to know if light acts insofar as it is perceived or if it is more directly determined by a tissular modification by the intermediary of the production of a hormone. It seems that light can act in a directly neuro-humoral fashion without psychological vision being necessary. In birds, light stimulates the retina and the interposed tissues up to the hypothalamus and hypophysis for if, in verifying experiments, light is directly shone on the hypophysis by means of a quartz rod, the stimulation of the gland and, through its intermediary, the gonads is obtained just as well as it is when light passes through the retina.[8] In the same way, we have seen that in insect societies, bio-social regulations take place equally through hormonal 'signals' (exchange of saliva) and perceptual signals in the strict sense (the contact of antennae). These transitional facts are interpretable if the organism is considered as a Form I and as a domain of primary consciousness. The hormone or inductive stimulus exercises a chemical action on the relevant tissue. It does not convey *information* in the strict sense: it *de-forms* or *trans-forms* the tissue in a certain manner. If this latter were a simple machine, or a step-by-step assemblage, in short, a zero-form, the chemical action would remain a pure, localised effect. But for the organism, being an absolute form in spatio-temporal self-possession, *trans-formation* is given in itself, it is subjectivity and primary consciousness as much as it is primitive form trans-formed.

The modes of action of hormones or inductors are probably very varied. Many have a direct role in metabolism: thyroid hormones seem necessary for cellular respiration, and their absence arrests growth. Since Thimann, we know that auxin is used in numerous reaction cycles which lead to the synthesis of new protoplasm. Many hormones act by catalysis; others are inhibitors or act on enzymes that are themselves catalysts or inhibitors.

But throughout this variety of modes, there is always the same fact of a transformation present to itself, which thus evokes a different morphological theme to the theme inherent in the non-transformed form. A perceptual signal stimulus, by transforming secondary consciousness – or the nervous area corresponding to this consciousness – likewise renders it capable of a new evocation. The animal that sees a sexual display is sexually excited. Perception in the strict sense is only a particular case of the action of a transformative, evocative stimulus, which would act, according to a macroscopic *patter* on a neural network, in the place of an action by its chemical form on an ordinary organic tissue.

This account naturally leads to the clumsy use of expressions borrowed from the experience of perceived stimulus in order to designate, in a semi-metaphorical way, the 'sensibility' of receptive tissue. There is a temptation to believe, consequently, that if the tissue, 'aware' of the action of the stimulus, nevertheless does not 'perceive' it, a mysterious property of 'sensibility', and then a new property to explain perception in the strict sense, must be presupposed. But everything becomes clear if it is properly understood that – here as it is throughout the rest of the domain of life and behaviour – everything that relates to Form I precedes and explains everything which relates to Form II. The 'sensibility' to stimulus of non-neural tissue is not, as it is in Cabanis' account,[9] a kind of mysterious additional property, a little like Anaxagoras's *Nous*, in the material presence of the tissue, but the form of the tissue itself, as 'absolute form', given in itself.[10] Perception in the strict sense, Signal Stimulus II, is not essentially different to Stimulus I. Simply put, stimulus is the trans-formation of a part of the organism that is organised into a network or modifiable screen, constituting an in-formation according to an exterior structure. *Consciousness II of information derives directly from Consciousness I of transformation.*

Light, in optical sensation, first acts by chemical action on rhodopsin[11] and on other substances. The stimuli of odours and tastes first certainly produce chemical effects. But what is essential is not the particular mode of these direct effects, which are often the origins of destruction or reversible lesions. What is essential is that they take place within a true form, not a zero-form, and that in turn, they possess themselves according to their structure, which thus becomes a perceived structure. Visual stimuli are macro-structures, by contrast with olfactory stimuli, in which the form of the molecule only acts according to prior effects of keys on the receptive cells, which bring them close – or less far away – from cases of Stimulus I. It is probably for this reason that odours are very often analogous in their mode of action to that of hormones or inductors, whose action they prolong into the domain of behaviour. In many animals, odours have a role in sexual adjustment, as in mating displays, but according to a mode of action which relates much more closely to sexual hormones. What is more, odours with a sexual role are often chemically related to sexual hormones.

TERRITORY AND THE VITAL DOMAIN

The close solidarity of Form I and Form II is again revealed in a series of phenomena which have, since von Uexküll, been studied a great deal: the organism is always in close relation to a vital domain whose forms are as characteristic as that of an animal that organises it as a den, as a burrow, as hunting terrain or terrain of refuge. Hédiger has described this external organisation of the vital domain as a *system of space-time* (*Raum-Zeit System*).[12] An animal does not generally use the whole of the objective space that it seems to have at its disposal: from this objective space (*Umgebung*), it cuts out a subjective space (*Umwelt*), but also a subjective time – in short, a system which composes everything which is significant for the animal, and which is organised by it or by the group to which it belongs.

> A vital domain is not homogenous; it is differentiated, generally composed of fixed points, of which shelter is the most important, the principal den, often completed with secondary shelters, points or zones of food supply, bathing, exhibition, excretion, points of reference, etc. These fixed points are united by a network of lines of communication, 'passes', whose width and sinuosity is often quite characteristic of the animal. Finally, the animal leaves its marks on the territory by scratching or organic secretion: urine, excrement or sometimes the products of specialised glands, like the preorbital glands of the antelope. Correlatively, the animal has rhythms of activity, temporal rhythms in the utilisation of its space.

The constitution of such territories naturally almost always, above all in the higher animals, presupposes a perceptual consciousness and behaviour regulated by objects. But the external morphogenesis of the vital domain is no less directly connected to organic morphology, perceptual consciousness only providing secondary adjustments. The burrow, for example, presupposes organs suitable for digging; marking presupposes specialised secretions or a special placement for the organs of excretion used for marking.[13] Temporal rhythms of activity depend even more clearly on specific physiological time. As Lambert and Teissier have suggested, homologies exist between animals in time as much as in space. Metabolic rhythms, the time of gestation and longevity all order the rhythm of behaviour just as anatomical forms order the structuration of the domain. The time of the mouse has the same relation to the time of the elephant as the mouse's tail has to the elephant's. Or, we must rather speak, precisely like Hédiger, of a *system of space-time* equally relevant for the organism and the vital domain of each species. It would be possible to go even further, and consider this system to be a veritable biological field or, according to an expression suggested by Hédiger, an 'active plane', in short, like a morpho-genetic theme ordering both the organic and extra-organic, internal and external circuits, biotope and psychotope, and which is to the interior what the skin is to the exterior.[14] A chick is made – or makes itself – in order to breathe, eat, reproduce, but also to peck at the external terrain: the 'Knowledge [*Savoir*] that is its *Umwelt*' comes to the same thing as the 'Knowledge which is its organism'. Organic movement includes the milieu. The *Umwelt* is in the position of a subordinated theme in the organic form before being differentiated as a distinct, extra-organic form, as territory.

> Extra-organic form is often homologous to the form of the body itself as if the formal theme were itself simply projected onto a much broader screen. This is the case with mollusc shells, the inverse projection of the extra-organic form onto the organic form able, exceptionally, to produce itself, such as in the hermit crab. This is also the case in the directing dances of bees, which transpose the long journey through the territory of nutrition into a reduced *system of space-time*, which is almost intra-organic, confronted with form and organic rhythm, or in any case intra-social. This is the case, finally, with the spider and its web, whose study has recently

been renewed by H. M. Peters. The spider's web constitutes a kind of 'territory' in Hédiger's sense, with a central shelter, lines of communication produced by the animal and materialising its rhythms and organic instincts. The Diadem spider's elliptical shape has the same form as the elliptical contour of its web; the web's central point is the spider's nest, as the central point of the spider is the ornament that gives it its name. When a spider's leg is amputated, this amputation is transmitted to the form of the web: the angles of the radial threads are correlatively changed.

Narrow behaviourist interpretations, which want to explain everything in terms of sequences of reflexes, miss the key point. In the structuration of a territory, behaviour in its totality dominates the perceptual indexes that guide it. It is not composed by the sum of automatic acts of obedience to perceived stimuli.

And the proof is in the fact that *the animal itself manufactures the sensory points of reference* of which it makes use. The markings made by mammals through the medium of the products of glands, or by urine, clearly have the characteristic of being 'voluntary' signals. Urine above all plays the role of a veritable hormone-odour or an inductor for the extra-organic form of the territory, thus confirming the interpretation of embryological inductors as 'organic signals' subordinated to a total plan. One part of the embryo *will carry* a chemical signal to another part. In marking its territory, the animal at once differentiates it like a tissue in formation, and individualises it, 'inhibiting' the formation of a territory by a stranger encroaching on its domain. It is, moreover, individualised in time, allowing for a continuity between the past and the future of what it recognises as its own. It would be possible to perform experiments on a territory analogous to those of experimental embryology. An artificial marking, for example, can produce a sort of 'secondary induction'. Inversely, many experiments in embryology can be interpreted in light of existing knowledge of territories. Secondary embryos in Spemann's experiments, for example, can be considered as products of 'false marking'. In P. N. Witt's celebrated experiments, making the *Zilla x notate* spider absorb neurotropic substances disturbed the external organisation of the web in the same way that the internal organisation of the embryo is disturbed when the embryologist intoxicates it at a precise moment. Certain of these disturbances – when the web, for example, is more regular than normal but also less well adapted to its milieu – seem to present a kind of denuded formative theme, cut off from sensory indexes.

Chapter 12
The Philosophy of Morphogenesis

Having compared the effort of identifying and tracking the mystery of morphogenesis to a criminal investigation, its conclusions can also be presented in the same fashion.

What appears clearly right away is the insufficiency of the notion of functioning borrowed from mechanist physics. A host of logical and experimental arguments can be used to prove the point that morphogenesis is irreducible to functioning, that is, to the setting in motion of a predetermined structure given in space.

THE IMPOSSIBILITY OF SPATIAL PRE-FORMS

This is, logically speaking, to misrecognise the position of the problem, to reduce formation – that is to say the *appearance* of a structure – to the *deployment* of an already existing structure. Functioning can only lead to the deterioration of that which functions.

Experimentally, the theory of preformation, which is the biological form of the theory of functioning, has been completely disproven. Even if, at the beginning of development, we look everywhere for hidden pre-forms – whose functioning would explain adult forms – we find nothing. Pre-forms are neither in chromosomes nor in genes. In fact, genes are distributed equally among all the divided cells in development. Let us admit, for the sake of argument, that genes could account for what Woodger[1] calls 'characterisation' (that is, specific characteristics) for the fact that the paw, tail or liver of a dog is not the paw, tail or liver of a cat. What it cannot account for, however, is 'organisation', the fact that these particular embryonic cells will become a liver, or a tail, rather than a paw. This is particularly clear for the two, four or eight cells that first result from the initial division of the egg. The nucleus has certainly been divided equally since each of the two or four initial cells will eventually yield a complete individual. It is not therefore the nucleus and its genes that explain the organisational differences between the right and the left half of the organism, or the front and the back, or the head and the tail. Incidentally, experiments with grafting show that the destination of a group of cells can be changed practically at will. What would have become a paw becomes a tail, and vice versa. Genetics cannot therefore explain embryogenesis.

Let us consider, again for the sake of argument, that the organisational pre-forms are found in the protoplasm. In many eggs, the protoplasm presents regional differences visible from the very beginning. The primer for bilateral symmetry, of the front and the back, the head and the tail can be recognised.[2] However, the objections against the pre-forms in the nucleus also hold for the pre-forms in the protoplasm. Grafting experiments modify not just cellular nuclei but the protoplasm as well, and yet the development is normal or normalised. The visible differences in the protoplasm appear, in general, to be due only to the presence of nutritional reserves. Their development is not always modified by their displacement through centrifugation. Should we then invoke invisible chemical differences? In the egg, whose development is 'mosaic', a chemical heterogeneity between the egg's parts can be found early on. In regulatory developments, though, a certain chemical homogeneity is conserved until gastrulation.[3] Now, regulatory development is, as we have seen, the fundamental case; mosaic development would simply signify a premature determination. If chemical heterogeneity is already the sign of a determination, that is to say, of the establishment of a formative process, how then could it be its cause?

Extraordinary difficulties are encountered if pre-forms of 'organisation' are located in the protoplasm, or pre-forms of 'characterisation' in the nucleus and the genes. In effect, during the course of evolution, the most general and fundamental organisational traits – which are today, according to neo-Darwinism, characters of kind or level – are due to first appear as mutations in a species. At this moment, according to the hypothesis, they have to depend on factors or pre-forms in the chromosomes. However – still according to the hypothesis – as characters of kind or level, they must depend on pre-forms in the protoplasm. Now, how can we possibly conceive of the transfer of these pre-forms from the nucleus to the protoplasm? Placental organisation, mammary glands, the mammalian instincts of nutritional lactation had to have appeared as genetic mutations. How could these genetic mutations themselves become the fundamental traits of their own organisation and embryogenesis?

THE INSUFFICIENCY OF RELATIONAL EPIGENESIST

Given that there are no pre-forms in space, there remains only to admit, for the sake of argument, a sort of epigenesis through spatial relations themselves. The increase in structural complexity, during formation, is explained by the external or internal relations of the egg in development – relations with its milieu at first, and then, after the first divisions, intercellular relations. This is the solution at which Woodger arrives.[4] A pure multiplication is not, by itself, a gain in complexity. However, a unified plurality of objects represents a level of organisation superior to that of each individual object that composes it. A forest can spring from a single tree, and yet the forest represents a superior level of organisation in the sense that, for example, the trees on the edge will have leafy branches reaching ground level while the trees in the centre will have leaves only at the canopy level. One fully-grown flower would thus be a sort of 'forest' of simple flowers, differentiated according to their place in the whole.

This 'relational' or 'social' explanation of morphogenesis, which is advanced today by Dupréel but also derivable from Gestaltist conceptions, has only a limited applicability.[5] We have shown that the situation, or the primary role of organic components in the whole, only acts as an evocative stimulus of capacity and not as its sufficient cause. This is already the case for the leaves of a tree in the forest and, even more so, in the differentiation of fully grown flowers – and yet even more so for the differentiation of cells in an animal. If the relational conception is true, the graft, in transplant experiments, should always act 'locally' and never 'originarily'. The determination of a tissue often depends in an indirect fashion on its location and its relations with another tissue. If, for example, a gastrulation towards the exterior is provoked, neutral differentiation does not take place due to a lack of good spatial relations, and the whole development is arrested. However, it would not be wise to take this as a ruling in favour of the 'relational' theory, which would be like explaining the painting of a scene by relating painter and canvas in space. The relational theory does not explain the organism's type, whose specific forms are maintained *in spite of* the milieu, and often in spite of incidental upheavals in their internal relations.

'Social' phenomena are real and of real importance in the whole domain of biology, but organic 'sociality' is irreducible to a simple spatial 'vicinity' produced by mechanically relational effects. Such a conception is only a return, in disguised form, to a theory of developmental functioning.

The efforts of biologists who cling to the idea of functioning are even more unjustifiable in light of its decades-old abandonment by physicists and chemists. What we have called the crisis of determinism is in fact the crisis – or rather, the abandonment without return – of the conception of physico-chemical phenomena as structures *first given in space, then put into motion*, and functioning according to *ready-made* connections. An atom is already in itself a process, a formative activity; it is not a functioning structure. The morphogenesis of an animal, as complicated as it is, nonetheless follows atomic and molecular 'morphogenesis', as the existence of the living molecules that are the virus show. And the morphogenesis of an animal reveals the same traits writ large that are found in the 'morphogenesis' of an atom, already in excess of functioning. There is today a true game of hide-and-seek between physicists and biologists. Biologists continue to make use of an out-dated chemistry; physicists, who, in general, have the most misconceptions about genetics, ignore the fact that embryogenesis is an active *formation*.

FORMATION AND CONSCIOUSNESS

Let us resume the investigation. Having recognised the insufficiency of functioning, one has to then search for the positive factor of morphogenesis. This factor appears, *as a first approximation*, as complimentary to functioning and takes on various aspects. They can be described as 'vertical themes', 'auto-conduction and auto-control', 'unitary behaviour', 'action according to an absolute surface', 'equipotentiality', 'mnemic melody', 'ability to react to a simple signal', etc. Under all of these aspects, the causal factor essentially represents an improvisation and a *creation* of liaisons, in contrast to the simple *play* of pre-given liaisons which characterises functioning. Finally, under all of these aspects the morphogenetic factor is revealed to be very close not to a mysterious 'vital principle' but to the immediate experience of consciousness. All these aspects (themes, auto-conduction, etc.) are at the same time aspects of consciousness. Consciousness is not a passive knowledge but the active unity of a behaviour or a perception. Consciousness *is* always a forming activity. It is always a dynamic effort of unification, without which 'behaviour' would be a pure collection of movements and perceptions a pure juxtaposition of physico-chemical effects able to be imitated by machines. It is therefore natural to suppose that morphogenesis is, to the contrary, always consciousness. This hypothesis, we must underline, does not consist in saying that consciousness *explains* morphogenesis; it rather asserts that consciousness and morphogenesis are one and the same.

It is nevertheless important to understand that psychological consciousness (whether human or animal) that perceives objects in the world and acts on the world is a morphogenesis in one particular organic domain, that of the nervous system – those veritable amoebic colonies that constitute the systems where liaisons are incessantly made and unmade according to themes derived from the broader theme of the organic but adapted to the outside world. This neural morphogenesis is then transposed, through the relays of organic machines, into movements in space. Naturally, though, this cerebral consciousness or morphogenesis is only a particular case, derived from organic consciousness and morphogenesis. An embryo in formation is a field of consciousness as much as it is an active cerebral sensori-motor area. It also improvises the new connections according to a theme; it is absolute surface and melody, like the cinema screen in *The Mystery of Picasso* on which each state of the painting serves as sign for another state.

Or better: it is because the embryo is the domain of primary consciousness that this embryonic part, consisting of nervous systems, can be the domain of perceptive consciousness and can facilitate organic behaviour by adjusting it to the extra-organic world. We walk and we see, and we manipulate objects because our cerebral nervous tissue is directly capable of modifying itself and of possessing itself absolutely in its thematic forms and deformations. *Our hands of flesh and bone are only the auxiliary machines of the 'absolute hand' of our cerebral cortex.* While the corporeal hand was being formed on the basis of the primordia of the embryonic limb bud, it was already 'absolute hand' – surface in possession of itself and sounding melody – independent of the cerebral hand that did not yet exist. However, in the adult organism – to the degree that it is alive, that is, capable of partially repairing and maintaining itself through nutrition and assimilation – it is no longer the 'absolute hand'. As tool-organ, it can only function like a set of tongs or pliers, and all control of its behaviour has been transferred to the cerebral hand. It is the cerebral hand that is the 'control' and consciousness of the active handling that facilitates the physical movements of the hand-organ.

We should not, however, be misled by this secondary dissociation. The embryo in development – to the degree that it has not already begun to function in accordance with deployed machines – is a complete field of consciousness. A nervous centre, an embryo or an embryonic area in formation, an amoeba or a unitary colony of amoebae such as *Dictyostelium* or even, let's add, a molecule in which the zones of individual indetermination have been reunited in a continuous network – all these domains are equally domains of consciousness, just as they are domains of morphogenesis.

Consciousness in morphogenesis is not a superimposed principle, a *deus ex machina*, or a 'ghost in the machine'. It is nothing other than form, or, rather, active formation, in its absolute existence.

Contrary to those theories inspired by (often poorly understood) Husserlian ideas, consciousness is neither always nor essentially 'consciousness of . . .', consciousness of a real or ideal object. The consciousness inherent in formation is not consciousness *of* a formation, either as light or as intention directed towards this formation. Primary consciousness is not 'consciousness *of* . . .' Only the consciousness of cerebral sensorial centres *qua* cerebral area becomes 'consciousness *of* . . .' Having been modulated by an exterior structure, consciousness envelops the existence of this structure or refers to an object through it and can as a result legitimately be called 'consciousness *of* the object'. The primary consciousness of formation, if this incorrect expression, involving a genitive that does not *refer*, absolutely must be used, is this formation *qua* formation that conforms to a general theme that dominates the constitutive elements. Consciousness *of* an habitual action is not the consciousness of performed movements (which would, on the contrary, disturb action); it is the active unification of constitutive elementary movements *according* to the theme of the action. Where primary organic consciousness is concerned, it is as illegitimate to employ the 'of' in Berkeley's sense as it is in Husserl's. Primary consciousness is neither consciousness *of* a perceiving Mind-subject nor the consciousness *of* an Object, whether real or ideal. Consciousness *is* any active formation in its absolute activity, and all formation *is* consciousness.

The viewer of the film *The Mystery of Picasso* has the illusion that the painting is painting itself since he does not see the painter behind the canvas. However, if we consider the painter's consciousness, the viewer's illusion corresponds to reality itself: the painting must form itself without a brush, held by a hand, being at work behind consciousness.

THE PHILOSOPHY OF 'VERTICALISM'

The investigation into morphogenesis cannot end here. Having identified formation and consciousness and jointly characterised them as auto-conduction and the improvisation of liaisons according to a theme, it is necessary to pass from description to an attempt at interpretation and to finally find, certainly not the cause of consciousness and formation – which would obviously make no sense – but rather their fundamental implication. At the beginning of this book, we invoked the metaphor of 'verticalism' to describe the general impression that first arises, given the facts of development. We saw very quickly that the metaphor has more than merely a descriptive value. Biological induction, the evocation through simple signals of formational competences, truly epigenetic appearance in space and time and specific complex structures all lead us to admit a non-geometric 'dimension' – a 'non-spatial' region in which the 'ideals' of specific forms subsist in a 'semantic' state, a state of significant themes analogous to the themes of an habitual act or an effort of memory or invention. At the same time, these ideals act dynamically on that which actualises them and are actively realised by it; in turn, these are adopted as *its* ideas. This particular composite of activity and passivity is characteristic of all consciousness. The cycles of mechanical auto-regulation merely 'symbolise' (in the Leibnizian sense) this characteristic and only represent it in a 'degenerated' state.

In any case, this conception suggests itself whenever we try to understand psychological consciousness. The studies of the Würzburg school, psychoanalysis, the psychology of instinct and above all – since the revolutionary conceptions of H. Jackson – the studies of aphasia have proven the reality of a semantic 'verticalism' and the central place of the trait [*tâche*], the theme or the dominant tendency in behavioural and intellectual formation, which are incomprehensible if we remain at the level of pure 'horizontal' associations. The brain serves as the 'control' – in the cybernetic sense of the word – for the movements of the organism. It is that which informs these movements and makes of them true behaviours, unified and guided. But while the 'control' of an automated machine is also a machine whose construction must be controlled by an engineer, the cerebral control operates directly because the brain, visible in space, is only the place where the non-mechanical feedback* to non-spatial ideals and themes is applied.

Let us observe the manner in which our actions on the outside world are regulated. The dynamic nature of our actions turns around the positing of a value-goal, and around diverse valencies, 'fixed' on intermediary-objects, inhibitive or supportive that are displaced as action progresses. These 'valorisations', inherent in conscious activity, correspond to the improvised modification of liaisons in the cerebral centres. Consciousness – or the formation of the current act – thus corresponds to the intersection of ideal themes and the organic machine in space. This 'dimension' of thematism is that which takes place in the adult and auto-conducting organism, the engineer controlling the 'control' of machines. According to J. C. Eccles,[6] the 'will' – it would be better to say the 'vertical' themes – transforms the spatio-temporal activity of nervous systems into a state of unstable closure by exercising already structured 'fields of influence'.[7] Alpha waves in no way represent, as it is sometimes suggested, a cerebral scanning*. It is rather that they cease as soon as a visual activity or mental calculation begins – which is in itself a return to a pure and autonomous functioning of rest of the nervous system, which escapes the control of 'vertical' themes.[8]

However, since psychological consciousness is only one particular case of organic consciousness, and the cerebral formations one particular case of organic formations, all recent discoveries in neuro-psychology must be valid, *mutatis mutandis*, for morphogenesis in general. All organic tissue in development is the site of the intersection between formative and regulative themes, and structures in space. All organic formation, like all cerebral activity, is controlled and regulated by non-mechanical feedback* *in accordance with a trans-spatial 'ideal'*. An in vitro cell culture into which a specific factor of differentiation is not introduced resembles a cerebral area at rest emitting alpha waves. However, in the developing embryo, all determined areas are put into circulation according to a theme that guides its differentiation by modifying its internal liaisons and displacing its valencies.

For the purposes of this account, we have spoken as if, by beginning with functioning and its recognised insufficiency, we had to look for a supplement which would transform functioning into a morphogenetic behaviour. In reality, of course, it is morphogenetic behaviour that is primary and functioning that is derivative in all true beings, as opposed to pure aggregates. Action, for modern physics, already indissolubly unifies time, space and energy; the cutting-out of an action in the time, space and energy of a particular system is always artificial and relative. The behaviour of an atom cannot be decomposed into a discrete functioning – itself already decomposed according to an absolute space-time – or into an x factor that modifies the functioning. The atom does not resemble the machines engaged by human activity *plus* something – on the contrary, phenomena and their laws make possible the existence of these very machines. Even less does the organism resemble a material machine, delivered passively into time and modified at each instant by an idea, an entelechy, beamed down from heaven. To recognise the dimension of a trans-spatial thematism indissolubly combined with spatio-temporal dimensions is not to accept the old dualism between body and soul, 'entelechy' and matter, idea and reality or vital principle and organic machine. The organism is not a machine *plus* a soul. Organic beings only subsist dynamically – in an incessant flux that, every few months, renews all its molecules. It is constant activity and the permanence of dynamism, not the permanence of a material reality informed retrospectively by an ideal form.

Having identified formation and consciousness, we must guard against the conception of consciousness as the attribute of a mind-substance, and against conceiving of thematism as the passive reflection of a static Platonic idea. Consciousness is neither a distinct ingredient, a sort of added phosphorescent substance, nor the attribute of a spiritual substance. Consciousness is nothing but the act, whether intelligent or instinctive, perennially engaged in the thematic organisation of sub-domains, themselves in the process of organisation. Cerebral consciousness, the active improvisation of formations in the nervous system, is at the root of the activity of organic consciousness that, for example, is ceaseless in its pursuit of the oxygenation of cerebral cells, or that actively maintains proteins in their form. The horse is not material organic tissue *plus* the Idea of Horse[9]. The horse is a horse because it 'horses'. It is not that, before passing through the 'blastula' stage, it is a pre-blastocoelic embryo *plus* the Idea of Gastrulation;[10] it has actively gastrulated, as actively as a bird migrates or nests. To conform to an idea, a mnemic or instinctive theme, is still to be active.

FORMATIVE ACTIVITY AND MEMORY

Following Whitehead, R. S. Lillie[11] has underlined the fact that while the activity that produces novelty appears to be the prerogative of consciousness, the constancy of things, the stable and conservative side of nature, appears to be physical.[12] Conscious existence is in the present and carries with it novelty and novel integrations. The past is what is left behind it as it advances into the future. The psychological, as Whitehead says, is always part of the creative advance of novelty. This conscious creation of novelty through integration always operates on a system that is pre-given and pre-structured by antecedent and subjacent activities, and it leaves in turn its structural imprint, its information – in the etymological sense of the word – on the system, which in this way develops according to an advancement of consciousness, while continuing to operate according to already acquired structures. This is to invert rather than to support the mechanical determinist view which asserts that the present actualisation – always action and always consciousness – is exclusively determined by the past. In fact, it is the past itself, or more precisely, previously acquired structures, which represents the traces left behind in the integrated sub-systems by the creative advancement of actualisation. Biological 'determination', far from being the result of a determinism – that is to say, the functioning of what already exists – is always prospective. It resembles the carrying out of a new construction plan; it inaugurates a new instalment of formation; it is the announcement that a new theme will be put into play. The time of pure functioning, in which the present proceeds from the past, is nothing but the conventionalised and deformed product of the time of conscious actualisation, inapplicable to organisms in formation. In a 'moment' [*tranche*] of conscious actualisation, there is no pure flowing of time from the past to the present but rather the circulation of an a-temporal theme in a domain, inaugurating an action that brings about a new spatio-temporal domain. This new domain appears to be continuous with the one that preceded it, but it does not flow from it like the sand in an hourglass.

It is necessary, nonetheless, to specify the nature of 'traces' and structurations left by creative advancement. These traces are, at a first approximation, of two very different types. Contrary to what R. S. Lillie[13] seems to believe, they are not solely physical and material – that is to say, they are not analogous to the traces left behind by an orator's voice on a vinyl record. They are also 'psychic', which is to say the actualisation of a theme or an idea produces recurrent effects on the non-spatial theme and modifies it according to an ascending action that passes from the actual to the trans-spatial. The orator who improvises a speech by actualising an idea produces physical effects that descend into the outside world; he produces a series of waves that can be recorded and conserved, due simply to the inertia of the vinyl or a magnetic metal. We can also suppose that the orator, after having spoken, takes notes so as to be able to eventually recite the improvised speech. Finally, we can suppose that the orator's nervous system, in its material structure, endures modifications analogous to those of the vinyl. But do conscious and creative advances have any effect other than material modifications? Such a thesis is unsustainable. Let us suppose that the orator would like to later repeat the speech that he has improvised. He consults his notes, which are, materially, only traces of ink on paper. They are nothing unless a conscious human being can understand and interpret them as signs. If in the meantime he suffers from agnosia, he will be incapable of using them. Can we say then that the interpretation of written signs depends on the sole cerebral traces of the orator, considered in themselves as a sort of writing or material recording left in the matter of the brain? But a material inscription, whether on cerebral tissue or paper, *cannot read itself.* Even the orator, having become aphasic, *tries* to speak; he still has the ideas whose actualisation is betrayed by the accidents that supervene on subordinate processes, psycho-motor schemes belonging to inferior levels that were developed through antecedent activity. He is not betrayed by a purely material confusion of purely material traces. An aphasic is not the same thing as a machine that prints words badly. His consciousness is an act directed by structuration, an act troubled less by the erasure of material traces than by the pathological state of his auxiliary psychic habits of structuration.

What gives rise to this misleading impression, and to the belief that material traces in themselves are sufficient, is the fact that it is possible to substitute the playing of the record for the presence of the orator. The banal dynamism of the phonograph's spring is the only thing required in addition to the structure of grooves on the record. Likewise, as Penfield's experiments have shown, the application of an electrode to the temporal lobe of certain epileptics seems sufficient to reactivate a sort of memory and an automatic recitation.[14] If active consciousness were really like this, then the mnemic act would be a simple amorphous force analogous to the force of the phonograph's spring. All it would seem to possess in terms of structuration would in reality be given to it by the structure of the traces, which it would simply put into motion once more.

In what other way could consciousness be capable of improvising and 'forming'? A conscious theme is not amorphous. It is structuring but is already and by itself structured in the sense that it includes a formal intention [*intention de forme*] that action only refines. Memory is essentially psychological, and the material traces can be nothing more than auxiliary. A mnemic theme is an ideal theme whose first actualisation has, through repetition, already been given a precise form. To claim the contrary is to return once again to the theory of pure functioning.

Suppose we were tempted to respond that after all it is not clear why we would transform what we claim to demonstrate into a postulate. Suppose we accept the notion of a pure cerebral functioning for psychological memory. And suppose we were to go as far as admitting that when the aphasic babbles, one part of the material brain reads another part of the same brain where the mnemic traces are printed, however badly. What will we have gained? Absolutely nothing – for if we pass from psychological memory to organic memory, we will no longer not be able to pretend that mnemic consciousness can be reduced to the functioning of material traces through banal and mechanical reactivation. If a banal induction of nervous circulations in the adult brain under a faradic current may seem sufficient to make it 'speak' its memories, this is already enough to rule out the claim that the formation of the brain, from the egg to the newborn, could be a similarly simple reactivation of structures readymade in the egg. We cannot claim today that the egg, with its genes and protoplasm, contains – like a kind of 'written plan', or like the record that only needs to be played – all the future forms of the adult organism and its nervous system. The whole of experimental embryology, and all the studies of instinct, prove that formational dynamic themes are truly formational and organisational. They do not simply deploy structures, make structures 'speak', since these structures do not yet exist, and since it is precisely the formational themes that give birth to them. The embryo constructs itself through the coordinated actualisation of a whole, non-spatial architecture of themes that is at once formational and already informed. The problems of embryonic formation are always essentially 'vertical', like the problems of aphasia. They manifest themselves through condensations, agglutinations, duplications, preservations, abnormal developmental arrests – in short, through a gruelling transition into the space of non-spatial themes.

THE NATURE OF MATERIAL TRACES

The duality of the mnemic effects of actualisation – material and psychological, spatial and trans-spatial – to which we had given provisory status is only apparent. Far from reducing everything to material traces, as the theory of functioning asserts, it is the material traces and spatial modifications that are reduced, in the final analysis, to thematic and trans-spatial modifications.

'Material' cerebral traces are supposed to be inscribed in the structure of cortical proteins and conserved by inertia against the passage of time. However, it can be shown, through a very simple calculation using results acquired through the method of isotopic marking,[15] that a protein molecule has an average lifespan of several days. Proteins are ceaselessly destroyed and reformed. What is more – as we know in the wake of modern chemistry – even over the same period, the subsistence of the molecule is in no way the mechanical inertia of a structure but an active persistence according to the rules of actualisation and spatialisation. We cannot therefore assimilate the traces eventually 'borne' by these molecules to the letters engraved in marble by a sculptor. Even if proteins reproduce, before disappearing, their exact double, traces included, the molecules bearing these traces are not equivalent to those fossils in which primitive organisms no longer subsist other than as petrified forms. Material traces in the ordinary sense of the word – grooves in the wax, or letters in marble – are only the secondary, solid effect encountered in our experience. Moreover, all material inertia is also only a secondary effect. The molecules of the wax or marble bearing 'traces' are themselves also active structurations. Just like organic proteins, their apparently inert structure depends on an actualisation, on a process always underway. It is primary memory, the trans-spatial subsistence of themes in activity, which creates the brute appearance of material inertia and the space-time of functioning as a secondary and statistical effect. It is actualisation, inventive or mnemic, and not the functioning of the past, which makes the present. The desire to explain the subsistence of forms through inertia is like wishing to explain the continuous activity of an atom through the inertia of a billiard ball.

In practical terms, we can speak of the material 'traces' left by an actualization on a material, conceived of as completely homogenous, when the trans-spatial theme of the formation is considered with respect to its *terminal* effects. The orator speaks before a tape recorder. As he speaks, according to a theme signifying the whole, he puts into play linguistic schemata and auxiliary motor schemata (already informed by preceding expressive efforts) in a cascade of improvised determinations analogous to the cascading determinations in embryogenesis. The thematic form of his intentions results, in the end, in modifications to a magnetic metal – in other words, in a modification of molecular relations. These modifications, for a modern chemist, are also modifications of atomic and inter-atomic 'activities', but roughly speaking and in practical terms, we can treat them as structural traces. The narrator's expressive effort has, along the way – in the vertical architecture of the trans-spatial – mnemic effects on linguistic schemata. When we speak, vibrations are produced in the air, but also and in the first instance, we learn to speak, creating partial ensembles better suited to expression in general. Terminal, so-called material modifications on the tape recorder or the nervous tissue are not fundamentally anything other than a kind of mnemic modification of the linguistic and psychological schemata. But because they occur at the end, we can for all intents and purposes take them as the spatial modification of a plastic matter.

Yet we must not be misled by this simplified manner of considering things and be taken in by the illusion of reducing the subsistence of the real to spatial inertia even though this inertia is only a limit idealisation. Embryology also results in physico-chemical phenomenon and seems to be reducible to them. But in this case, the illusion is more difficult to maintain – even though the blindness of biases can uphold it – due to the enormous gap between the point of departure and the point of arrival. It is difficult to convince those who want to reduce the memory of the orator to the material traces in his brain. The brain is so much more complicated than we are capable of imagining it to be. It should not be difficult to argue that the memory by which an egg becomes a human being is irreducible to material traces in the egg and that it implies a whole architecture of trans-spatial themes in which the egg, and then the embryo, are only the spatial (or quasi-spatial) fulcrum, modified throughout their evocation.

THE PYRAMID OF FORMS

At the end of this investigation into morphogenesis, we therefore find it necessary to admit a kind of non-geometric dimension containing formational themes. 'Verticalism' is not simply a metaphor. These themes regulate the incessant activity that makes life. Just as psychological consciousness is always an effort according to an ideal sense, an effort which is translated by psycho-physiological ensembles themselves never completely imitable to the mechanical ensembles of automatism, formative consciousness is always an effort according to themes, making it more stereotypical and mnemic in character but without it becoming any less ideal and trans-spatial.

An economy of hypotheses is a good thing. The perseverance of biologists in explaining formation by physico-chemical phenomena is admirable. But the virtue of economy can be pushed too far and at times reveals only a lack of imagination. He who persists in making four equilateral triangles with six matchsticks laid flat on a table without thinking about arranging them in a tetrahedron has also achieved an economical, if misguided, solution.

In all of the domains in which complex, organic or para-organic forms are found, it is remarkable that the pyramid of forms seems set down on its apex. Written language is composed from twenty-four letters and a few dozen signs; the most complex sentences and speech always come down to a few dozen fundamental sounds; music rests on a handful of notes. In

253 the nervous system, the most complicated actions and shrewdest manoeuvres always come down to the same few muscular commands. In the same way, the most elaborate calculator comes down to a play of elementary electrical impulses that substitute 0 for 1, and 1 for 0. More generally, the unbelievable variety of phenomena in the entire universe is always reduced in the end to the displacement and rearrangement of the same elementary particles – electrons, neutrons, neutrinos. It is truly inconceivable that the whole pyramid is accounted for by its apex, by the movements of particles in space, and that the greatest masterpieces, in nature as well as in art, are only 'an alphabet in disorder'.[16] For the pyramid to hold, we require a proper consciousness of forms.

MORPHOGENESIS AND REASON

To explain all forms (whether of type I, II or III) by the zero-form, in a fashion more or less renewed by Democritus – which is to say, by the fundamental disorder of atoms or elements – is in every way excluded by modern science, which only recognises derivative phenomena in the zero-form, in statistical molecular agitation and in the equilibriums and laws of classical physics. The subsistence of forms I, II or III can only depend on a direct relation with a domain of order.

If we try to understand organic morphogenesis *in general terms* – leaving aside for the moment the detail of scientific explication, like someone listening to a speaker without also thinking about the sounds that are being uttered, or like the user of a machine who tries to roughly understand the role of its parts without following in detail its processes of realisation – we clearly grasp *a reason* in forms. For example, we see clearly that every organism must use sources of energy like a machine. We see clearly the reason why respiration,
254 in a higher-order multicellular organism, requires a more complex system of channels than the respiration of a protozoon; we see why the respiratory system of an insect can be simpler than that of a mammal, and why the heart of a mouse or a sparrow must beat faster than the heart of an elephant. We see clearly the reasons for the organs of photosynthesis in plants. We see the reason why a plant can and must have a mode of growth very different from that of an animal, with solidified parts which no longer develop, and specialised parts that ensure on-going growth. We also see the general reason which presides over the diverse systems of organic or inter-organic coordination, in cellular societies and animal or human societies. In short, organic forms are intelligible in their general technique, which is troubled by the same problems and often finds the same solutions as the more lucid and more self-aware hu-

man technique. Long before we had formulated a definition of cybernetics, we had come to realise that organic techniques could inspire industrial techniques, and that the progress of industrial techniques could allow for a better comprehension of organic techniques. All forms, whether of type I, II or III, appear to depend on the same Reason.

But what is mysterious is the way in which diverse kinds of beings could attain this Reason. In order to advance their technology and perfect the forms of their industrial machinery, civilised human societies are required to create research departments and organise scientific research. Where are the research departments and the CNRS[17] of organisms? Yet even before human laboratories, organisms discovered flight, electric batteries, calculators, ultrasound and so on.

The same applies to the problem of organic invention in general, as it does to the problems of vision, locomotion and manipulation. As a result of a strange anthropomorphic naïveté, we believe that technical invention is natural and explicable if it is due to a human being, if there is a professional inventor endowed with a good brain and working in a subsidised laboratory. A technical invention in an organism, without a professional scientist or a research laboratory, appears mysterious and paradoxical, and we see no other reasonable solution than to attribute it entirely to chance mutation.

This is to simply forget that the human brain which invents itself is first of all only an organic tissue, a network of cells, and that *every human and social deployment of invention is only auxiliary and accessory*. In the human invention of the radar or flight, everything fundamentally rests upon the auto-conduction of some cells of grey matter in which, according to a research theme, the instructions for assembly must have been combined in themselves.

'How can a cellular colony, without a brain, invent the rational and technical dispositions of the organism?' The question is naïve. What is the brain if not a colony or cellular network? The human who stands amazed before the organic inventions of an amoeba colony or an embryonic tissue simply forgets that his own inventions are themselves organic inventions and cortical cellular formations, subsequently transposed.

We leave to pseudo-rationalists the assertion that it is superstitious to believe in organic finality and that finalist action can only be conceived in human psychology and thanks to the human brain. The broadly speaking rational character of morphogenesis is explained by the connecting up of every organic domain with the world of trans-spatial themes. Forms I are just as connected up with the themes as Forms II and Forms III. Or rather, forms II and III are only connected up with the themes because they are particular cases of Form I. The human is only conscious, intelligent and inventive because all living individuality is conscious, intelligent and inventive.

THE HOMOGENEITY OF INTELLIGENCE

There is a fundamental homogeneity of consciousness, intelligence, finality and the capacities for generalisation and abstraction in all organisms, according to a sense. These features belong in an essential fashion to all true forms. Each and every organic individuality, in the broadest sense of the word, is not only an absolute surface in possession of itself, a field of consciousness, but also an inventive intelligence.

The psychologists who fabricate so-called IQ tests run into serious difficulty every time they want to utilise them for culturally non-homogenous groups.

> In applying, for example, the first versions of the Binet-Simon test to rural and urban boys and girls belonging to different social classes, the test's topics appeared to advantage boys of bourgeois parents, while disadvantaging city children. The same mental exercise, depending on whether it involves a marble or a doll, can appear easier for a boy or a girl. A test that asks what the word 'sonata' means is easier for bourgeois children than it is for a working-class child. In order to remedy this inconvenience, the tests are 'balanced', equalising them until they no longer favour one particular group[18] – but then it becomes impossible to draw any conclusions about the intellectual equality or inequality of the tested groups. The same scores, for example, between boys and girls simply prove that the tests are well-balanced. Unequal scores do not necessarily prove that boys and girls are intellectually unequal but perhaps just that the tests have been insufficiently balanced. It is more difficult yet again to attempt to make the tests, even those that are not language-based, what is called culture free*. Drawings that represent a violin, a mechanical pencil, a pocket-knife or a telephone would naturally be indecipherable for Melanesian children. Only topics supposed to be common across all cultures and trialed in diverse cultures can be used (cross cultured tests*) – but the simple use of paper and crayons, or even the simple presentation of abstract marks without practical signification in a testing environment, favours or handicaps certain cultures.

It is easy to see how serious this situation is, not only with respect to the significance given to IQ tests but to the very idea of intellectual difference. And in fact, if we follow this to its conclusion, we rediscover the same fundamental difficulty when we conduct experiments on the psyches of various animals. Broadly speaking, a chimpanzee appears to be more intelligent than a dog, and a dog more intelligent than a hen. But it would be necessary to run 'instinct-free'* tests to actually decide. The chimpanzee has a hand, along with an instinct to hold on to branches, along with its own instinctive 'stick-age'. This handling of the stick gives humans the impression of intelligence, above all because it recalls a human gesture. The dog's paw is incapable, for good reason, of such a performance – but does this prove that the dog is less intelligent, or only that the dog, in its 'organic culture', in its instinctive 'ethology', applies its intelligence at other points?

We can even go as far as the amoeba, which would be even more handicapped than the young Melanesian or by the pen-and-paper test. Would this be an absence of intelligence, though, or the lack of a certain 'acquired content' in its mode of organic culture? Is it more intelligent to walk with legs and eat with a mouth than it is to succeed at eating and walking without legs or a mouth, with only the appropriate deformations of a protoplasm? We might say that it is more intelligent to have developed, in the course of evolution, a handy set of legs and a mouth. But is this chance or skill? Chance and luck, which are absurd to invoke as replacements for consciousness and organic intelligence, are capable of explaining the *unequal satisfaction* of organic intelligence as it comes to grips with different milieus and circumstances. Ethnologists hesitate to link intellectual inequality to the 'inequalities' of human cultures since they perceive all too well the differences in directions of application. Likewise, the belief in the greater intelligence of a particular human being is, in most cases, pure class prejudice, this intelligence simply being applied to a broader scale or a more specific material. A cabinet minister does not have to make a greater intellectual effort in balancing a budget of hundreds of billions than a mayor does in regulating the spending of his town. A manual labourer would have been able to become a laboratory scientist if he had applied his intelligence to different objects. It is not in principle any more difficult to find a conclusion to a syllogism when it bears on atoms or electrons than when it bears on marbles, even though it is a fact, as experience shows, that a subject little familiar with a certain 'material' of reasoning allows himself to be disconcerted by it. We all have, in the same way, what could be called a 'species prejudice', a biological prejudice. The least civilised, including many primitive humans beings, consider, with a wisdom worthy of Montaigne, the animals they hunt to be beings as cunning as humans but in possession of different habits. Not only Montaigne, but also the psychologists of instinct who today engage in 'comparative ethology' and who consider cross-sections of animal and human cultures on the same level [*plan*] are our precursors here.[19]

What gives the thesis of the homogeneity of intelligence a falsely paradoxical, and even purely fictional appearance, is, of course, that which interests the researchers as it does those who employ humans or animals – namely, the genuine, actual or quasi-actual capacities of individuals and species. Now, however, the worker who would have been able to become a laboratory scientist cannot do so any longer. The chimpanzee is capable of performances of which cats and dogs are not. It is *practically* impossible to disassociate intelligent activity from its habits of application. But for the general problem of formation – our problem – this paradox is truth itself. As C. T. Morgan emphasises, the capacity of generalisation, of reaction 'to what appears similar' can be observed in animals located right at the bottom of the phylogenetic ladder, and 'in this respect, there has been no essential change throughout phylogenesis'.[20]

Given Spearman's *g* factor – that is, given the characteristic capacity of intelligence and cerebral consciousness to pass from given terms to the relation that unites them, or from a term and a given relation to a second term united to the first by the relation[21] – we can discern a general organic capacity that we can dub the '*gamma* factor', which is not only 'noegenetic' but 'morphogenetic', and which acts according to the same laws. What is reproduction, regeneration and the characteristic equipotentiality of all life if not the capacity to 'generalise', to act according to the similar or the thematic rather than according to pure causes? Since, as we have noted, even the reproduction of a virus or a protein cannot be a mechanical moulding, it must rather be an 'eduction of correlates', indissolubly both morpho- and noegenesis.[22]

We have thus only been able to rediscover our fundamental conclusion, and the identification of formation and consciousness. We must not forget Spearman's two principles – 'eduction of relations', and 'eduction of correlates' – themselves dependent on a first principle which he rightly calls the 'principle of consciousness', or 'the principle of the apprehension of experience':[23] 'All lived experience tends to immediately evoke a knowledge [*connaissance*] of its character, and an experiencing "I"'. This is to say that consciousness and life are one.

METAPHYSICAL EPIGENESIS

To recognise the homogeneity of consciousness or intelligence throughout the domain of life is not to add an adventurous metaphysics to a study that desired to stay as close as possible to scientific evidence but rather to gain the means to respond to a last and apparently serious difficulty. By rejecting the false idea of functioning we reject all preformism. But by invoking a transspatial thematism are we not led to replace a mechanical preformism with a metaphysical preformism, simply placing the models of form outside space instead of looking for them within it? The response can be drawn from human experience since it is homogenous with all organic consciousness. The experience of technology or art clearly shows that morphogenesis through human intervention is guided by ideas, by glimpses of the possible or harrowing experiences of the impossible, while being in no way copied from a model. The inventor knows in general terms what he wants – what Claparède aptly calls guided invention – but he cannot read the details of the form of what is to be created in himself or in the heavens and must engage in trial and error. By analogy with the radio, we desired the television, glimpsing its possibility and suspecting which lines of research would be involved, but its model existed nowhere – no more in the Platonic heavens than in our space.

The guide of consciousness or active intelligence is not an Engineer or a transcendent Architect.

It is precisely the human experience of invention that forecloses the possibility of deriving any form of anthropomorphism from the principle of creation and organic morphogenesis. The prophet or the guilty sinner freely imagines a kind of Super-man, hanging over them, who speaks to or threatens them. But the inventor or artist, the creator of forms, believes in a standard of success and even inspiration, while nonetheless knowing very well that this inspiration is not, in an event, whispered to them from the wings. It is he and no one else who must correct, retouch, eliminate the faults of the work and laboriously draw near to the idea which he wants to incarnate in it. And it is also he who must profit from strokes of good fortune by noticing and preserving them. This fundamental approach remains unchanged when we pass from human to organic invention, from psychological noegenesis to organic morphogenesis. The organism too forms itself amidst risk and peril; it is not formed. The differences between them, as considerable as they are, do not bear on the essential. They concern, on the one hand, the more mnemic character of organogenesis, such that it resembles the filling in of a crossword puzzle. The puzzle's author, rediscovering the grid and the list of questions created earlier, applies them now anew in order to resolve his own set of questions. On the other hand, they concern what the organism fabricates directly and does not – like *homo faber* – have to transfer its 'fabrication', through the cerebral relays, into an extra-organic matter. The living being is at once the agent and the 'material' of its own action. It is identical to brain tissue which would not have had to play the role of a first relay in an extra-organic realisation, and which is self-sufficient. The living being forms itself directly in accordance with a theme, without the theme first having to become an idea-image or represented model.

The difference between morphogenesis and noegenesis is in the end superficial. The living being forms itself just as the psychological idea forms itself in us, if not in the way the idea is subsequently realised with our hands. The true human experience of invention, true invention, that of the idea as such, frees itself from analysis and takes place through a direct actualisation. To cite the poet D. H. Lawrence,

> Even an artist knows that his work was never in his mind.
> He could never have thought it before it happened.[24]

Invention is guided by a theme, it does not proceed by chance. But to conceive this trans-spatial theme as a model to be copied – in invention or morphogenesis – would be to be duped by the completely secondary and particular character of human invention.

Morphogenesis is neither the work of a copyist nor a pure active force. Correlatively, its directive Logos is not the patternmaker of a grand couturier or the creator of mechanical robots. It is in fact a non-spatial order, an unformulated yet effective ideal, a guide to activity indissociable from this activity itself. It does not keep for itself all real being, leaving forms to be mere copies; neither is it a pure illusory Nothing. The reality of organisms and of actual beings presupposes a non-Parmenidean being. An action, or an authentic formation, escapes from the Parmenidean dilemma of being and non-being. Being, opposed to non-being, cannot characterise an 'active being' since an 'active being' is by definition striving to be but *is* not. If it purely and simply was, it would not act. Being, opposed to non-being, can no longer characterise the directing ideal, the theme of an as yet unformed form. If it were, it would no longer have the need for an active actualisation. Only the set {theme → form} is. To separate one term from the other is to condemn them both to vanish. Active, thematic formation alone is. Its conventional decomposition into 'pure theme' and 'pure form' leaves nothing but two shadows. To cite Lawrence once again in response,

> Even the mind of God can only imagine
> Those things that have become themselves.

Notes

INTRODUCTION

1. Arthur Stanley Eddington, *The Philosophy of Physical Science* (New York: The Macmillan Company, 1939), 15–52.
2. Bertrand Russell, *Introduction to Mathematical Philosophy* (New York: Dover, 1993 [1919]), 61.
3. TN. This is perhaps a reference to Paul Scarron's *Virgil Travesti*, a partial-verse parody of Virgil's *Aeneid*.
4. TN. Ruyer is referring to the eighth section of Plotinus's *Ennead III*, §2. Speaking of the productiveness of nature, he writes, 'It is clear, I suppose, to everyone that there are no hands here or feet, and no instrument either acquired or of natural growth, but there is need of matter on which nature can work and which it forms' (Plotinus, *Ennead* III, 1–9, trans. A. H. Armstrong [Cambridge, MA: Harvard University Press, 1967], 363).

CHAPTER 1:
Verticalism and Thematism

1. TN. Ruyer is referring here to Albert Dalcq's *L'oeuf et son dynamism organisateur* (Paris: Albin Michel, 1941).
2. TN. The chordal or axial process is what gives rise to the gelatinous inner core of the spinal column.
3. TN. This remarkable term describes the dyeing (using various and increasingly more complex techniques) of embryonic cells such that their eventual 'normal' outcome can be predicted. This technique, first deployed by Walter Vogl in 1929, allows for the production of 'fate maps': coloured, pictorial representations of what each part of the embryo will, 'under normal conditions', become.
4. Cf. J. S. Wilkie, *The Science of Mind and Brain* (London: Hutchinson's University Library, 1953), 126.
5. Wilkie, *Science of Mind and Brain*, chap. 3.
6. TN. An electrical current between 50 and 100Hz.
7. TN. A tank designed (both at the level of chemical composition and of mechanisms for stirring and modifying temperature) for the precipitation of crystals.

8. Arnold Gesell, *L'embryologie du comportement* (Paris: Presses universitaires de France, 1953).

9. See the reproductions of Dr. Hooker's cinemicrophotographs in Gesell, *L'embryologie*.

10. Gesell, *L'embryologie*, 40.

11. TN. 'Sketched out' translates *ébauché*, a verb form of the noun *ébauche* whose English correlate is 'primordium', as we indicated in 'Note on the Translation'.

12. These are C. C. Speidel's observations.

13. TN. Again, as indicated in 'Notes on the Translation', Ruyer is playing on the two meanings of *ébauche*: as an artistic 'sketch' and as a primitive developmental state.

14. TN. Ruyer is referring to a device, first invented in the seventeenth century, which allows a drawer to copy in enlarged or reduced form any shape using a simple mechanism of four jointed arms.

15. H. J. Jordan, 'Indéterminisme vital et le dynamism des structures causales', in *Recherches Philosophique* 2, ed. Alexandre Koyré, Henri-Charles Puech and Albert Spaier (Paris: Boivin, 1933), 30.

16. Jordan, 'Indéterminisme vital', 29.

17. Jordan, 'Indéterminisme vital', 30.

18. TN. Ruyer does not provide a reference for this passage, which we have translated directly from his French. Its provenance, moreover, has proven difficult to establish. An English translation of Driesch's famous presentation of the results of these experiments conducted in Trieste over two months in 1891 can be found in Hans Driesch, 'The Potency of the First Two Cleavage Cells in Echinoderm Development: Experimental Production of Partial and Double Formations', in *Foundations of Experimental Embryology*, ed. Benjamin H. Willier and Jane M. Oppenheimer (New York: Hafner Press, 1964), 38–50. The passage closest to the one that Ruyer presents reads as follows:

> The first time I was fortunate enough to make the observations described above, I awaited in excitement the picture which was to present itself in my dishes the next day. I must confess that the idea of a free-swimming hemisphere or a half gastrula with its archenteron open lengthwise seemed rather extraordinary. I thought the formations would probably die. Instead, the next morning I found in their respective dishes typical, actively swimming blastulae of half size. (46)

It is difficult to say whether Ruyer has simply reconstructed this passage or is citing some other passage of Driesch from elsewhere.

19. C. H. Waddington, *Principles of Embryology* (London: George Allen & Unwin, 1956), 17, 23. But Waddington does not forbid himself use of this 'wild card'.

20. Cf. J. Needham, *Order and Life* (Cambridge: Cambridge University Press, 2015 [1936]), 157; and John Tyler Bonner, *Morphogenesis: An Essay on Development* (Princeton, NJ: Princeton University Press, 1952), 45.

21. TN. See note 18 above.

22. Emmanuel Fauré-Fremiet, 'Symétrie et polarité chez les ciliés bi- ou multicomposites', *Biological Bulletin* 79 (1945), 106–50.

23. R. G. Harrison, 'Relations of Symmetry in the Developing Embryo', cited in Bonner, *Morphogenesis*, 219.

24. TN. A spicule is a needle-shaped component of these organisms in which it functions in large numbers as a de facto skeleton.

25. D'Arcy Wentworth Thompson, *On Growth and Form* (London: Dover, 1942), 679.

26. Bonner, *Morphogenesis*, 43.

27. TN. A catenary (*chaînette*, both from the Latin *cateneria*) is the quasi-parabolic shape formed by a chain, rope or cable held only by its two ends.

28. TN. Capable of modulating its internal activity in relation to a range of unpredictable environmental inputs, the homeostat was an early computer built by William Ross Ashby in 1948. His ambitions for it are outlined in *A Design for a Brain* (New York: Wiley & Sons, 1952).

29. TN. This term has now been effectively abandoned in cell biology and medicine for the more general 'antibody', but it describes a co-adaptive physiological agent that brings together two others; for instance, the serum of someone suffering from syphilis and an antigen introduced into it.

30. Raymond Ruyer, *La cybernétique et l'origine de l'information* (Paris: Flammarion, 1954).

31. TN. 'Accommodation' is a technical term that describes the capacity of an eye to adjust its focal length, thereby keeping an object in focus as the object moves. The eye of the snail is not capable of doing this.

32. An example elaborated by H. J. Jordan.

33. We have done so elsewhere; see Raymond Ruyer, *Neofinalism*, trans. Alyosha Edlebi (Minneapolis: University of Minnesota Press, 2016), chaps. 16–17; and Raymond Ruyer, 'Les postulats du sélectionnisme', *Revue philosophique de la France et de l'etranger* 146 (1956), 318–53.

34. Cf. E. Schrödinger, *What Is Life?* (Cambridge: Cambridge University Press, 1948), 65.

35. H. G. Bray and K. White, 'Organisms as Physicochemical Machines', *New Biology* 16, no. 70 (1954), 74. TN. For some reason, Ruyer indicates that Ilya Prigione was also a co-author of this piece, which is untrue.

36. John von Neumann, 'The General and Logical Theory of Automata', in *The World of Mathematics*, vol. 4 (London: Dover, 2003), 2070–98. We have outlined and discussed von Neumann's reasoning at length in Ruyer, 'Les postulats du sélectionnisme'.

37. Von Neumann, 'General and Logical Theory', 2095.

38. Cf. J. B. S. Haldane, *The Biochemistry of Genetics* (New York: The Macmillan Company, 1954).

39. Gavin de Beer, *Embryos and Ancestors* (Oxford: Clarendon Press, 1940).

40. Walter Garstang, 'The Theory of Recapitulation: A Critical Re-statement of the Biogenetic Law', *Zoological Journal of the Linnean Society* 32 (1921), 82.

41. Julian Huxley, 'Evolution as a Process', in *Evolution as a Process*, ed. Julian Huxley, A. C. Hardy and E. B. Ford (London: George Allen & Unwin, 1954), 12.

42. Cf. E. W. F. Tomlin, *Living and Knowing* (New York: Faber & Faber, 1955), 90.

43. TN. The French inventor Louis Blériot created the aircraft dubbed the Blériot XI in the early years of the twentieth century. The Super Constellation was a plane designed and built for commercial flight by the Lockheed company.

CHAPTER 2:
From the Molecule to the Organism

1. Jean Rostand, *Les grands courants de la biologie* (Paris: Gallimard, 1951), 162.

2. TN. In contrast to the practice elaborated in the early nineteenth century, a developed formula presents not just the relative proportions of atoms composing the molecule but also the specific position of each atom. This allows for distinctions between different isomers with the same components (for instance, CH_2Cl-CH_2-CH_3 and CH_3-$CHCl$-CH_3).

3. TN. Ruyer is invoking Genesis 2:6–7.

4. Albert Frey-Wyssling, *Submicroscopic Morphology of Protoplasm*, trans. May Hollander (New York: Elsevier Publishing Company, 1953), 371. TN. We have added the ellipsis here to indicate that Ruyer has skipped over a number of sentences.

5. TN. Ruyer's text here reads '*comme des feuilles attachés à la branche et non comme des feuilles détachées et tournoyant dans l'air*'. This is in fact a fairly plain paraphrase of the English rendering of Frey-Wyssling's claim. Writing of the structures of living things, he writes – in a passage, part of which Ruyer has clearly been glossing in earlier sentences and continues to invoke in what follows, 'They are not intermingled by mere laws of chance and Brownian molecular movement; the fact is rather that they arrange themselves into a delicate, very plastic and flexible pattern, actuated, as it were, by a purposeful, co-ordinative impulse. No more than leaves, blown by autumnal winds from the twig and fluttering helplessly in the air, are able to assimilate for the parent tree, can independent, ambulant, reactive molecules take part in any organised work' (*Submicroscopic Morphology*, 373).

6. TN. The second half of Ruyer's version of the citation reads, '*Les centres actifs du réseau protoplasmique s'arrangent selon un pattern souple qui semble guide par une impulsion finaliste coordinatrice*'. Once again, then, he silently contracts and modifies the passage cited in the previous note.

7. Frey-Wyssling, *Submicroscopic Morphology*, 374.

8. TN. Eddington certainly writes that 'there is in a human being some portion of the brain, perhaps a mere speck of brain-matter, perhaps an extensive region, in which the physical effects of his volitions begin' (*The Philosophy of Physical Science* [New York: The Macmillan Company, 1939], 182). However, in his earlier Gifford lectures that appear in *Nature of the Physical World* (Cambridge: Cambridge Scholars Press, 2014), Eddington explicitly rejects the hypothesis – already familiar from Descartes' *Passions of the Soul* – that 'the mind operates through two or three key-atoms in the brain', stating that it is 'too desperate a way of escape for us' (309). It would appear, then, that Ruyer is using the phrase 'key-atom' as a metonym for Eddington's conviction that some portion of matter is directly affected by an ideal cause.

9. René Poirier, 'Henri Poincaré et le problème de la valeur de la science', *Revue philosophique de la France et de l'etranger* 74, nos. 10–12 (October–December 1954), 485–513.

10. J. B. S. Haldane, 'The Origins of Life', *New Biology* 16, no. 12 (1954), 20.

11. TN. A polar globule or polar body is a cell made during the process of ovulation, but for the most part and unlike the ovum, it is not viable and cannot be fertilised.

12. J. D. Bernal, 'The Origin of Life', *New Biology* 16, no. 12 (1954), 18.

13. Cf. G. Bachelard, *La matérialisme rationnel* (Paris: Presses universitaires de France, 1953), in particular chapters 4 and 5.

14. Bachelard, *La matérialisme rationnel*, 146.

15. TN. The chemist Auguste Kekulé was the first to describe benzene with alternating single and double bonds in a static representation:

16. Cited in Bachelard, *La matérialisme rationnel*, 66.

17. Pierre Morand, *Aux confins de la vie* (Paris: Masson, 1955), 74.

18. TN. T2, T4 and T6 are bacteriophages (viruses that affect bacteria) originally extracted from *E. coli*.

19. TN. In this particular case, the gas is given off by rotting fruit.

20. TN. The cosmologist George Gamow was an important early participant in debates on quantum physics. A failed early attempt to defect from the Soviet Union found Gamow and his wife crossing the Black Sea in a kayak stocked with only chocolate and brandy.

21. Cf. H. Blum, *Time's Arrow and Evolution* (Princeton, NJ: Princeton University Press, 1951), chaps. 7 and 8.

22. TN. We have of course chosen to render *se comporter* in this way, and against the broader use here of 'to behave' and its cognates, in order to avoid the very different meaning of the English reflexive verb 'to behave oneself'.

23. TN. Ruyer is referring to John Tyler Bonner (1920–), whose work deals predominantly with certain photosynthetic algae (diatoms) whose cell walls have the unusual feature of being composed of silicon dioxide.

24. John Tyler Bonner, *Morphogenesis: An Essay on Development* (London: Scribner, 1963), 90.

25. Cf. Eugene Willis Gudger, 'Fishes That Rank Themselves Like Soldiers on Parade', *Zoologica: Scientific Contributions of the New York Zoological Society* 34 (1949), 99–103.

26. Mary Field, J. V. Durden and F. Percy Smith, *See How They Grow* (Harmondsworth: Penguin, 1952), 151.

27. Frey-Wyssling, *Submicroscopic Morphology*, 186–88.

28. TN. ATP, or adenosine triphosphate, is the basic molecular component of DNA and is characterised by its capacity to store and transmit energy.

29. Cf. Blum, *Time's Arrow and Evolution*, 102ff.

30. Georg Christoph Lichtenberg, *Aphorismes*, trans. Marthe Robert (Paris: Club Français du Livre, 1947), 118.

31. TN. This phrase – commonly used in English to describe the process Ruyer is describing – translates the uncommon French term *épictèse*, from the Greek *épi* (in addition, supplementary) and *ktésis* (acquisition). The word appears to have no direct English correlate.

32. TN. This term transliterates Ruyer's *amboception*, an uncommon word in both French and English also found in the work of Jacques Lacan. It derives from the Latin *ambo* (both) + *ceptor* (receptive agent). It invokes, in the current context, a relationship of complementary coupling between the various organs and macro-processes of the organism.

33. TN. See chap. 1 n. 28.

34. TN. It is likely that Ruyer is invoking Gamow's *The Birth and Death of the Sun* (New York: Viking Press, 1940) in which Eddington's classic post-Helmholtzian hypothesis of the intra-stellar equilibrium, presented two decades earlier in *The Internal Constitution of the Stars*, is discussed rather than Eddington's work itself.

35. François Meyer, *Problématique de l'évolution* (Paris: Presses universitaires de France, 1954), 183ff.

36. Ramond Ruyer, *La cybernétique et l'origine de l'information* (Paris: Flammarion, 1954), chap. 5.

37. Meyer, *Problématique de l'évolution*, 189.

38. This is Barron's comparison – see Blum, *Time's Arrow and Evolution*, 110, 119.

39. Meyer, *Problématique de l'évolution*, 190.

CHAPTER 3:
Internal Reproduction

1. Cf. H. Blum, *Time's Arrow and Evolution* (Princeton, NJ: Princeton University Press, 1951), 132.

2. Adolf Portmann, *Animal Forms and Patterns* (New York: Schocken Books, 1952), 38.

3. Hans Driesch, *La philosophie de l'organisme*, trans. M. Kollmann (Paris: Rivière, 1921), 101ff.

4. Cf. Joseph Henry Woodger, 'The Concept of "Organism" and the Relation between Embryology and Genetics', *Quarterly Review of Biology* 6 (1931), 202.

5. TN. Here, 'regulative' specifically invokes the kind of equipotential cell differentiation that concerns Ruyer in which the 'destiny' of a cell can be transformed in a variety of ways. In contemporary embryology, it is opposed to 'mosaic' development, which is the *partes extra partes* process that Ruyer seeks to reconceptualise on the basis of a radical expansion of 'regulation'.

6. Bonner, *Cells and Societies* (Princeton, NJ: Princeton University Press, 1955), 101.
7. Bonner, *Cells and Societies*, 104.
8. TN. Ruyer's list includes both French pronouns for 'you' (*tu* and *vous*).
9. TN. Throughout this next paragraph, the italicised passages translate Ruyer's recurrent use of capitalised pronouns: 'Quand je me souviens, ou quand il me souvient, ou quand je rêve'.

CHAPTER 4:
The Fragmentation and Socialisation of Development

1. TN. A releaser (for instance, a releaser melody sung by a bird) is a stimulus that evokes a certain instinctive behaviour in its receiver. Tinbergen's famous study of the male stickleback concerned its responses to anything coloured red – notably, the bellies of other male sticklebacks – during mating season.
2. Nickolaas Tinbergen, *Social Behaviour in Animals, with Special Reference to Vertebrates* (London: Chapman & Hall, 1990), 114.
3. Eugène Dupréel, *La pragmatologie* (Brussels: Les Editions du Parthenon, 1955).
4. Talcott Parson.
5. Konrad Lorenz, *Les animaux, ces inconnus*, trans. C. Jouoan (Paris: Editions de Paris, 1953), 135.
6. Melville Herskovits, *Man and His Works* (New York: Knopf, 1949), 212. TN. As he does elsewhere, Ruyer's citation here is a contraction of Herskovits's actual prose.
7. Karl von Frisch, 'Lernvermögen und Ergebunden Tradition im Leben der Bienen', in *L'instinct dans le comportement des animaux et de l'homme* (Paris: Masson et Cie, 1956), 20.
8. Alfred North Whitehead, *Adventures of Ideas* (New York: Free Press, 1967), 205.
9. TN. The Abbey of Thélème, a kind of ideal community governed only by the rule 'Do what you will', appears in the first volume of François Rabelais' utopian work *Gargantua and Pantagruel*. Ruyer's invocation of the abbey here draws attention to Rabelais' derivation of natural goodwill in human beings from a certain picture of biological life.
10. Tinbergen, *Social Behaviour in Animals*, 114.
11. D. W. Morley, *The Ant World* (New York: Penguin, 1995), 10, 76.
12. Morley, *The Ant World*, 76.
13. Morley, *The Ant World*, 76.
14. TN. According to Fourier's analysis of human psychology – which he modelled on the force of gravity – there are three basic organisational passions: *cabaliste* (intriguing, dissident); *papillonne* (alternating and contrasting) and *composite* (the need for two pleasures at once).
15. E. P. Deleurance, 'Analyse du comportement bâtisseur chez les Polistes (Hyménoptères Vespides): L'activité batisseuse d'origine "interne"', in *L'instinct dans le comportement des animaux*, 105–50.

CHAPTER 5:
Signal Stimuli

1. Edward Stuart Russell, *The Behaviour of Animals: An Introduction to Its Study* (London: Edward Arnold & Co., 1938), 112.
2. Russell, *The Behaviour of Animals*, 112–13.
3. TN. The existence of a class of inductive or 'messenger' chemical compounds (organisins) was hypothesised by the French biologist Theodore Lender in 1955 in the context of experiments on the relationship between the regeneration of the brain and eyes in flatworms.
4. Cited in John Tyler Bonner, *Cells and Societies* (Princeton, NJ: Princeton University Press, 1955), 99.
5. What is fundamental 'effectivity' in non-statistical physics? We will put this question to the side. What we know about this today (concerning the role of the energy of exchange of photons and bonding electrons or pi mesons) allows us to assert that they do not, in any case, have anything in common with the transmission of movement through contact described by the mechanical models of the nineteenth century (which, incidentally, skirted the problem of causal effectiveness).
6. TN. A pantograph is a drawing mechanism typically composed of four interconnected arms that allows its user to copy and enlarge whatever it traces.
7. TN. Nematocysts are features of the organisms of the phylum Cnidaria (jellyfish, for instance). They consist of small pods that open onto the surface of the body and that, when triggered, open up and propel a barbed hook (often poisonous) towards whatever acted as a trigger.
8. TN. A now familiar example would be a neurotransmitter like dopamine, which is the 'key' for the 'lock' constituted by various receptors found in the central nervous system.
9. TN. Ruyer is referring to the class of locks that include some mechanism that obstructs or hides access to the keyhole. The simplest example would be a hinged shutter that would have to be held aside in order for the key to be inserted. The analogy is meant to imply that the apertures through which the automatons supposedly take in information from the environment are in fact closed in advance and that 'new' information can only be an iteration of what the automaton already knows.
10. TN. In 1943, and in the wake of Turing's account of computation, Warren McCulloch and Walter Pitts advanced an early version of the neural network model of the brain in which neurons were conceived of as binary machines structured according to a set of input-output relations.
11. TN. *Abboné*: someone with their own access to the telephone network; akin today to having a mobile phone plan.
12. Raymond Ruyer, *La cybernétique et l'origine de l'information* (Paris: Flammarion, 1954).
13. Cf. O. Köhler's experiments (Tinbergen, *Etude de l'instinct*). TN. Ruyer is probably referring to Wolfgang Köhler's experiments in the learning behaviour of apes.
14. TN. Karl von Frisch is famous for his experiments on communication behaviour – taking the form of 'dancing' – in honey bees.

15. TN. A trigger. Konrad Lorenz proposes a generalised theory of the trigger-stimulus, which he argues is what provokes the coming into activity of an instinct.

16. Benoit Mandelbrot, 'Structure formelle des textes et communication', *Word* 10 (1954), 1–27.

17. Those of Vowles on ants, and Birnkow (1954) for ladybirds. Cf. Karl von Frisch, 'Lernvermögen und Ergebunden Tradition im Leben der Bienen', in *L'instinct dans le comportement des animaux et de l'homme* (Paris: Masson et Cie, 1956), 361.

18. TN. Precise English versions of this phrase do not appear in Lorenz's classic *King Solomon's Ring*. In his discussion of jackdaw cries, however, Lorenz does speak of an unconscious expression of mood:

> All expressions of animal emotions, for instance, the 'Kia' and 'Kiaw' note of the jackdaw, are therefore not comparable to our spoken language, but only to those expressions such as yawning, wrinkling the brow and smiling, which are expressed unconsciously as innate actions and also understood by a corresponding inborn mechanism [. . .] The jackdaw or goose merely gives unconscious expression to its inward mood and the 'Kia' or 'Kiaw', or the warning sound escapes the bird involuntarily; when in a certain mood, it must utter the corresponding sound, whether or not there is anybody there to hear it (Konrad Lorenz, *King Solomon's Ring*, trans. Marjorie Kerr Wilson [London: Routledge, 2004], 75, 79).

19. TN. The collective noun that Ruyer uses (*bandes*) would traditionally be rendered into English as 'clattering'.

20. This is Tinbergen's observation.

21. Cf. von Frisch, 'Lernvermögen und Ergebunden Tradition', 361.

22. Johannes Holtfreter, 'A Study of the Mechanism of Gastrulation', *Journal of Experimental Zoology* 95, no. 2 (1944), 171–212.

23. Cf. M. Fontaine, 'Analyse expérimentale de l'instinct migrateur des poissons', in *L'instinct dans le comportement des animaux*, 151–67.

24. TN. The psychologist Joseph Banks Reine was the first to engage in controlled studies (broadly and loosely speaking) of 'extra-sensory perception' in the 1930s, with results published in a 1934 book of the same name. Ruyer's reference to Bergson here is most likely to *Creative Evolution*, published in 1907, where the latter deals with biological topics familiar to Ruyer and claims at a number of points that intuition is sympathy.

25. TN. Narziss Ach was an experimental psychologist and member of the Würzburg School best known for his work on concept formation and the unconscious bias introduced by the framing of expectations.

26. Abraham Moles, *Théorie de l'information et perception esthétique* (Paris: Flammarion, 1957), chap. 1.

27. TN. '*On ne dit plus un véhicule automobile, mais une auto, ou une voiture*'.

28. TN. 'Pigeon milk', or 'crop milk' as it is more commonly known, is a high-protein, fat- and nutrient-rich excretion produced by a range of birds (male, female, or both, depending on the species) for the feeding of their young. It is excreted from the crop or ingluvies, a pocket at the base of the oesophagus. The secretion is triggered by pressure from the beaks of the feeding birds.

29. TN. As this set of examples indicates, the term has a different semantic range in French and English. In French, *une operation* can be a business deal but also a miraculous intervention (what in English would be called a 'work', as in 'a work of the Lord').

30. TN. Birds belonging to the order Galliformes, ground-nesting birds that are most often reared for domestic consumption.

31. Russell, *The Behaviour of Animals*, 25.

CHAPTER 6:
Competence

1. TN. Mechanical watches of the vintage Ruyer is referring to sometimes came to a halt through a combination of friction and wear, and shaking these watches could often restart them.

2. TN. H. J. Watt's experiments early in the twentieth century concerned the nature of associations between words – the 'task' that Ruyer refers to, for instance, 'Associate a word that indicates a part of the entity in the given word' – and the ways in which these associations could be modified by various pre-existing and induced mental states. Watt's main conclusion was that the task itself determines the nature of the associations that it frames, thereby displacing the idea of a judging subjectivity that supervenes in the act of association.

3. Joseph Needham, *Biochemistry and Morphogenesis* (Cambridge: Cambridge University Press, 1942), 112.

4. Needham, *Biochemistry and Morphogenesis*, 263.

5. John Tyler Bonner, *Cells and Societies* (Princeton, NJ: Princeton University Press, 1955), 158.

6. J. Benoît, 'Etats physiologiques et instinct de reproduction chez des oiseaux', in *L'instinct dans le comportement des animaux et de l'homme* (Paris: Masson et Cie, 1956), 177–260.

7. Needham, *Biochemistry and Morphogenesis*, 141.

8. Gavin de Beer 'Embryology and Evolution', *Journal of Philosophical Studies* 5, no. 19 (1930), 482–84.

9. TN. Ruyer is referring to the taxonomic orders of frogs and salamanders respectively here, and buccal tissue is from the mouth. Oscar Schotte (1895–1988) was a Russian developmental biologist. Despite misspelling his name (Ruyer spells it 'Schotté'), Ruyer's various invocations of the *Triton* (newts) are indebted to Schotte's pioneering doctoral study of their capacity for regeneration.

10. TN. The spadefoot toad and the moor frog.

11. Reported in Needham, *Biochemistry and Morphogenesis*, 348.

12. TN. A mole salamander.

13. TN. *Necturus* is a species of salamander including *Necturus maculosus* (the mudpuppy) and *Necturus punctatus* (the dwarf waterdog).

14. Cited in John Tyler Bonner, *Morphogenesis: An Essay on Development* (Princeton, NJ: Princeton University Press, 1952), 202.

CHAPTER 7:
Autonomous Procedures and Regulated Behaviour

1. Cf. Adolf Portmann, *Animal Forms and Patterns: A Study of the Appearance of Animals* (New York: Schocken Books, 1952), 131.
2. Jakob von Uexküll, *Animal Forms and Patterns: A Study of Appearances of Animals*, trans. H. Czech (New York: Schocken Books, 1967), 128.
3. Konrad Lorenz, 'La théorie objectiviste de l'instinct des animaux et de l'homme', in *L'instinct dans le comportement des animaux et de l'homme* (Paris: Masson et Cie, 1956), 53.
4. TN. Weiss worked on newts in particular.
5. TN. Ringer's solution, named for its inventor the physiologist Sidney Ringer, is a commonly used saline solution containing sodium chloride, potassium chloride, calcium chloride and sodium bicarbonate in the ratios in which they appear in the body of the animal under study.
6. TN. 'Double assurance' here refers to the multiple determination of morphogenetic outcomes. Ruyer is almost certainly invoking Joseph Needham's *Chemical Embryology* (Cambridge: Cambridge University Press, 1931) – especially since he makes specific reference to the same species of frog and makes the same points about amphibian ocular lens development. Needham defines double assurance in the following terms: 'Cells only become what they do under the influence of as many as three or four contributing causes. The double assurance principle may correspond to the factors of safety which appear in structural engineering, so that, if one process goes wrong, the embryo can still manage to complete its development with the aid of the others' (*Chemical Embryology*, 579).
7. TN. This example is actually reported in B. P. Wiesner and N. M. Sheard, *Maternal Behaviour in the Rat* (Oxford: Oliver & Boyd, 1933), 7. Russell himself cites this text rather than reporting his own experience of the dog in *Instinctive Behaviour* (London: Edward Arnold, 1934), 129: 'Whenever she finds a rag she starts pawing and trying to cover up with her muzzle some imaginary object'.
8. TN. The vitelline sac is more commonly known as the membrane around an egg yolk – which, as Ruyer indicates, is not found in the human umbilical vesicle.
9. Daniel Lehrman, 'On the Organisation of Maternal Behaviour', in *L'instinct dans le comportement des animaux*, 475–520. TN. On crop, and crop milk, see chapter 5, note 28.
10. Lehrman, 'On the Organisation of Maternal Behaviour', 12. TN. Ruyer briefly describes the characteristic approach of the American school of comparative psychology, which emphasised strict laboratory control in experiments and the importance of behavioural context (i.e., 'nurture') in 'Three Modes of Thematic Activity' above.
11. Cf. Nikolaas Tinbergen, *L'etude de l'instinct*, trans. B. de Zelicourt and F. Bourlieue (Paris: Payot, 1953), 106.
12. Gabriele d'Annunzio lost an eye in a flying accident in 1916 while deployed in World War I. The self-styled *arcangelo mutilado* nevertheless went on to compose

the prose poem *Notturno* on short, thin strips of paper while he healed, his whole head swathed in bandages.

13. TN. Here and below, Ruyer plays on the French term for these sections (*disques*), which is also the term for vinyl records that are 'spun' (*faire tourne*) on a record player.

14. Cf. Karl Lashley, 'The Problem of Serial Order in Behaviour', in *Cerebral Mechanisms in Behaviour* (New York: Wiley, 1951), 112–36.

15. This is G. Dumas' example. Transcriptions of the speech of aphasics provide a profusion of example. TN. 'J'ai étyré un pousan!' – 'J'ai épousé un tyran'. A simple English example: 'Time for me to shake a tower' for 'Time for me to take a shower'.

16. TN. Ruyer's French, which conveys the point somewhat more effectively, reads: '*Il l'a ti jamais attrapé, le gendarme, son voleur?*'

17. Cf. Andre Ombredane, *L'aphasie et l'elaboration de la pensée explicite* (Paris: Presses universitaires de France, 1951)

18. TN. See note 13 above.

19. Lashley, The Problem of Serial Order', 127. TN. We have introduced an ellipsis in order to indicate a short section of the passage that Ruyer has left out.

20. TN. Ruyer's example phrases in this paragraph are *Il les regarde*, *Il les regzrde*, *Il les regardes* and *Il les regardent*. The material layout of the keyboard commonly used in France (AZERT) differs from the Anglo-American (QWERTY) format, hence our modification of the first example.

21. TN. These three categories of plurals refer respectively to cases like 'cat/cats', 'medium/media' and 'ranch/ranches'. Here are Ruyer's French examples: -*x* can pluralise some nouns, for example, *lieu/lieux* (place/places); and -*ent* can pluralise a large class of verbs, for example, *ils épuisent* (they are exhausted), as opposed to *il épuise* (he's exhausted). In English, the addition of a suffixed letter to pluralise a verb is much rarer, hence our use of two forms of pluralisation for nouns above.

22. Cf. René Poirier, 'Le problème de l'âme et du corps', *Société Française de Philosophie* 48, no. 4 (1954), 97–158. But R. Poirier completely grasps the difficulties of the hypothesis.

23. Lorenz, 'La théorie objectiviste de l'instinct', 68.

24. TN. Ruyer is referring to a well-known 1930 study by Zing-Yang Kuo, 'The Genesis of the Cat's Response to the Rat', *Journal of Comparative Psychology* 11 (1930), 1–35. Ruyer thus wrongly identifies the rodent in question.

25. TN. In fact, Cournot's maxim was '*S'il y a toujours un hasard pour faire échouer une entreprise, ce n'est pas par hasard que cette entreprise échoue* [If there's always a chance an enterprise will fail, it's not by chance that the enterprise fails]'.

26. Anna Anastasi, *Differential Psychology: Individual and Group Differences in Behaviour* (New York: Macmillan, 1966), 66.

27. TN. The taxonomic order containing frogs and toads.

28. Heini Hédiger, *Observations sur la psychologie animale dans les parcs du Congo Belge* (Brussels: Institut des Parcs Nationaux du Congo Belge, 1951), 86.

29. Cf. Edward A. Armstrong, *La vie amoureuse des oiseaux*, trans. Jane Fillion (Paris: Albin Michel, 1952), chaps. 4, 5 and *passim*.

CHAPTER 8:
Open Formations and Markovian Jargon

1. TN. 'Secteur abrité'. Ruyer is perhaps playing here with the more common use of this phrase, which designates the non-traded sector of an economy: all economic activity available to the nation's occupants alone (healthcare, for instance). The womb, in this analogy, refers to the sphere of economic activity free from direct interference by the exterior milieu and its 'cataclysms'.

2. Cf. Georges Théodoule Guilbaud, *La cybernétique* (Paris: Presses universitaires de France, 1954), 76.

3. Cf. Pierre Guiraud, *Les charactères statistique du vocabulaire* (Paris: Larousse, 1956).

4. TN. Titus Livius (or Livy) was an historian, most well-known for his voluminous (and only extant) work, a history of Rome.

5. TN. 'Angel', 'perfume', 'heart' or 'breast', 'ecstasy', 'blood' and 'demon'.

6. TN. 'Azure', 'nude', 'blank' or 'virgin', 'gold', 'dream' and 'pure'.

7. Cited by Frederik Jacobus Johannes Buytendijk, 'L'instinct d'alimentation de l'expérience chez crapauds', *Archives Néerlandaises physiologie de l'homme et des animaux* 2 (1918), 221.

8. TN. In 1952, Albert Ducrocq built Calliope (named for the Greek muse of poetry), a small computer that illuminated a light bulb at random intervals, thereby generating a sequence of random bits of information that could be translated into words.

9. TN. Ruyer's formalisation of Zipf's law – that there exists a functional dependency between the frequency of a word and the number of meanings the word possesses – is less than standard. Today, it is more often stated in a form equivalent to $\mu \alpha f^\delta$, where μ is the number of meanings a word has, f is the frequency of the word and the constant $\delta = \frac{1}{2}$.

10. Pierre Vendryès, *Vie et probabilités* (Paris: Albin Michel, 1942), 333.

11. TN. This term, along with the example of the Parisian taxi driver that Ruyer cites in what follows, is also taken from Vendryès.

12. Benoit Mandelbrot, 'Structure formelle des textes et communication', *Word* 10 (1954), 21.

13. Cited by Melville J. Herskovits, *Les bases de l'anthropologie culturelle* (Paris: Payot, 1952), 100.

14. TN. Max Weber, *The Protestant Ethic and the Spirit of Capitalism*, trans. Talcott Parsons (New York, 1958), 91. Ruyer does not provide a bibliographic reference for this passage, but it – along with the invocation of a 'partial elective affinity' – is drawn from Merleau-Ponty's *Adventures of the Dialectic*, cited in what follows.

15. Maurice Merleau-Ponty, *Adventures of the Dialectic*, trans. Joseph Bien (Evanston, IL: Northwestern University Press, 1973), 24, translation modified.

16. François Meyer, *Problématique de l'évolution* (Paris: Presses universitaires de France, 1954).

17. Cf. Gilles Gaston Granger, *Méthodologie économique* (Paris: Presses universitaires de France, 1955).

18. TN. Ruyer is referring to the argument presented in John von Neumann and Oskar Morgenstern, *Theory of Games and Economic Behaviour* (Princeton, NJ: Princeton University Press, 2004 [1944]), the foundational work of game theory.

19. TN. Unlike symbiosis, commensalism names a unidirectional or non-mutual relation between two species such that only one species benefits; unlike parasitism, the relationship is not deleterious for the passive party. A species of sea-sponge, for instance, provides habitation for small marine life without the latter contributing anything to the on-going existence of the sponge.

20. Nikolaas Tinbergen, *L'etude de l'instinct*, trans. B. de Zelicourt and F. Bourlieue (Paris: Payot, 1953), 233.

21. M. Autuori, 'La fondation des societes chez les fourmis champignonnistes du genre *Atta* (Hym. Formicidae)', in *L'instinct dans le comportement des animaux et de l'homme* (Paris: Masson et Cie, 1956), 77–104.

22. D. W. Morley, *The Ant World* (London: Penguin, 1953), 92.

23. Julian Huxley, *Fourmis et termites*, trans. William Perrenoud (Brussels: Office de publicité, 1955), chap. 7.

24. TN. Though Ruyer does not indicate this himself, the quoted passage is from the end of the second stanza of Paul Verlaine's 'Art poétique':

> Il faut aussi que tu n'ailles point
> Choisir tes mots sans quelque méprise :
> Rien de plus cher que la chanson grise
> Où l'Indécis au Précis se joint.

25. Richard Goldschmidt, *Physiological Genetics* (New York: McGraw Hill, 1938).

26. Cf. J. B. S. Haldane, *The Biochemistry of Genetics* (New York: Macmillan, 1954), 111–24.

27. TN. Both of these latter instances occur because the caterpillar or ant that comes behind another follows the trail of biochemical signals left by the one in front, a sequential relation that can become circular rather than linear, leading to the phenomena of the death spiral and the 'ant mill' that Ruyer invokes here.

CHAPTER 9:
'Crossword' Formations

1. Arnold Gesell, *L'embryologie du comportement* (Paris: Presses universitaires de France, 1953), 185.

2. Gesell, *L'embryologie*, 202.

3. TN. The CGT (*Conditions générales de transport*) is the European code of conduct governing international rail travel.

4. TN. Ruyer is referring here to Jacques Bénigne Bossuet, a theologian and historian, and more specifically to his most influential work, the 1681 *Discours sur l'histoire universelle*. With respect to Leibniz, the key text is the famous allegory of

the pyramid that closes the *Theodicy*, which considers the place of Sextus Tarquin's life in the context of the best of all possible worlds.

5. Nickolaas Tinbergen, *Social Behaviour in Animals, with Special Reference to Vertebrates* (London: Chapman & Hall, 1990).

6. Cf. C. H. Waddington, *Principles of Embryology* (London: George Allen & Unwin, 1956), 455.

7. Joseph Needham, *Biochemistry and Morphogenesis* (Cambridge: Cambridge University Press, 1942), 231; and Etienne Wolff, *La science des monstres* (Paris: Gallimard, 1948), 35. Wolff has show that, contrary to Needham's view, double monsters are unequally composed.

CHAPTER 10:
'Spectacle-Spectator' Complexes

1. Cited by J. B. S. Haldane, 'Les aspects physico-chimiques des instincts', in *L'instinct dans le comportement des animaux et de l'homme* (Paris: Masson et Cie, 1956).

2. TN. Ruyer gives no direct reference. The most well-known context in which Berkeley advances this claim is near the start of *A Treatise Concerning the Principles of Human Knowledge*: 'For as to what is said of the absolute existence of unthinking things without any relation to their being perceived, that seems perfectly unintelligible. Their *esse* is *percipe*, nor is there any possibility that they should have any existence, out of the minds or thinking things which perceive them' (George Berkeley, *A Treatise Concerning the Principles of Human Knowledge*, ed. Kenneth Winkler [Indianapolis, IN: Hackett, 1982], part 1, §3, 24).

3. Julian Huxley, *Evolution in Action* (New York: Harper, 1953), 91.

4. TN. It is worth noting that Ruyer is referring to earlier televisions in which the image was 'written' on the back of a phosphorous-coated glass surface by cathode ray tubes (CRTs, or 'electron guns') firing differently charged electrons through a sequence of filters. What was seen on the screen was thus indeed 'cast'.

5. A quite similar appearance is achieved on the lower surface of the wings of the owl butterfly (*Caligo*) through what is necessarily a completely different procedure since the wing's surface is made up of overlapping scales like tiles on a roof and not contiguous barbules. In the case of the owl butterfly, furthermore, the designs are completely independent of the underlying structure and, despite the slenderness of the wings, they are completely independent of the decorations on the upper face, on which complex, iris-like effects are produced.

6. TN. '*Se* comporte et *se* perçoit'. It is not quite possible to render Ruyer's French entirely here. While the French *comportement* is straightforwardly translated as 'behaviour', the reflexive verb form *se comporter* can mean both 'to behave' or 'to behave oneself' – neither of which is appropriate in this context where the reflexive *se* is meant to indicate the auto-affective relationship that characterises all living forms.

7. TN. Explaining 'the obscure by the even more obscure'.

CHAPTER 11:
Forms I, II and III

1. TN. This last remark, which seems to contradict the general thrust of Ruyer's distinction between the three ranks of form, should be understood to mean that the organic body can be grasped as an assembly of Forms I in a particular structure, aside from the 'organic techniques' of the membrane and tubes discussed in the next section, and aside from the receptive capacities of the specialised zones of perception discussed above.

2. Gaston Bachelard, *La matérialisme rationnel* (Paris: Presses universitaires de France, 1953), 146.

3. TN. Ruyer refers here to John Hughlings Jackson whose work on aphasics was concerned with the fact that they were at times capable of using the vocal function for other purposes (e.g., singing) even when they could not speak.

4. TN. This number is now thought to be 15.

5. TN. '*C. pallidus* [. . .] possesses a remarkable, indeed a unique, habit. When a woodpecker has excavated in a branch for an insect, it inserts its long tongue into the crack to get the insect out. *C. pallidus* lacks the long tongue, but achieves the same result in a different way. Having excavated, it picks up a cactus spine or twig, one or two inches long, and holding it lengthwise in its beak, pokes it up the crack, dropping the twig to seize the insect as it emerges [. . .] This remarkable habit [. . .] is one of the few recorded uses of tools in birds' (David Lack, *Darwin's Finches* [Cambridge: Cambridge University Press, 1983], 58–59).

6. Nikolaas Tinbergen, *L'etude de l'instinct*, trans. B. de Zelicourt and F. Bourlieue (Paris: Payot, 1953), 230. TN. This should be read as Ruyer's gloss of Tinbergen rather than an accurate repetition of the passage in question.

7. J. Benoît, 'Etats physiologiques et instinct de reproduction chez des Oiseaux', in *L'instinct dans le comportement des animaux et de l'homme* (Paris: Masson et Cie, 1956), 177–260.

8. Benoît, 'Etats physiologiques', 204–5.

9. TN. Pierre Jean Georges Cabanis was a (now mostly ignored) precursor to modern neuroscience and an enduring influence on Schopenhauer. He not only advanced an early form of neural functionalism but theorised the notion of 'nervous energy', an analogue to electricity produced and moved around the brain.

10. M. Maurice Vernet has reprised a thesis related to Cabanis' but in a much more subtle form (*La sensibilité organique* [Paris: Flammarion, 1948]).

11. TN. Ruyer uses the more colloquial 'visual purple' [*pourple rétinien*] here.

12. Cf. Heini Hédiger, *Observations sur la psychologie animale dans les parcs du Congo Belge* (Brussels: Institut des Parcs Nationaux du Congo Belge, 1951); and *Les animaux sauvages en captivité: Introduction à la biologie des jardins zoologiques* (Paris: Payot, 1953).

13. TN. As in male hippopotamuses (Cf. F. Bourlière, *Vie et moeurs des mammifères* [Paris: Payot, 1951], 78).

14. Heini Hédiger, 'Instinkt und Territorium', in *L'instinct dans le comportement des animaux et de l'homme* (Paris: Masson et Cie, 1956), 532.

CHAPTER 12:
The Philosophy of Morphogenesis

1. Joseph Henry Woodger, *Biological Principles* (London: Kegan Paul, 1929), 349.

2. TN. A DNA primer is the brief initial sequence of nucleic acids that DNA replication requires in order to begin.

3. Cf. J. Needham, *Biochemistry and Morphogenesis* (Cambridge: Cambridge University Press, 1942), 139. TN. Gastrulation is an early moment in embryogenesis during which the initial surface formation of the embryo (the blastoderm, see note 10 below) further develops into a three-layered structure (ectoderm, mesoderm and endoderm).

4. Woodger, *Biological Principles*, 351.

5. Cf. Eugène Dupréel, *La pragmatologie* (Brussels: Parthénon, 1955).

6. John Carew Eccles, *The Neuro-physiological Basis of Man* (Oxford: Oxford University Press, 1953). We do not follow Eccles when he invokes Rhine's suspect experiments and his 'telekinesis' to materialise this will as a real force, applying it to neurons as it is applied to any other object such as dice or playing cards. An idea, a theme of action, is dynamic in the present through the systemic unity that it gives birth to in the organic domain in which it is the trans-spatial correlative, but it does not come to bear on energy any more than it does on matter. Consciousness is dynamic without being a particular form of energy. Its dynamism is borrowed from the individualities that it unifies. It is indivisible energy which is born in the attenuation of individuality of the constituents of the system. What appears to the physicist as bonding through energy exchange is nothing other than an elementary field of consciousness.

7. Eccles, *Neuro-physiological Basis of Man*, 277.

8. Cf. also A. F. Adrian, *The Physical Background of Perception* (Oxford: Clarendon Press, 1947); and Alfred Fessard, 'Mechanisms of Nervous Integration and Conscious Experience', in *Brain Mechanisms and Consciousness*, ed. Jean-François Delafresneye (Oxford: Blackwell, 1954), 229.

9. TN. 'Horse' and 'horses' are translations of Ruyer's rather archaic *Cabaleité* and *cabaliser*.

10. TN. The blastula stage is an early moment in embryogenesis during which the cells polarise into an exterior (blastoderm) and interior (embryoblast).

11. R. S. Lillie, *General Biology and the Philosophy of Organism* (Chicago: University of Chicago Press, 1945), *passim*, especially chap. 12. We have also emphasised this point – see Raymond Ruyer, *Éléments de la psycho-biologie* (Paris: Presses universitaires de France, 1946), 109ff.

12. Lillie, *General Biology*, 161.

13. Lillie, *General Biology*, 96, 164.

14. Penfield's observations have often been quite 'embellished' in second-hand accounts. Penfield notes that an 'applied stimulation to what seems to be the same point of the cortex can produce an entirely different memory'. The memories evoked are, most importantly, thematic and not stereotypical. It is in this way alone that a patient expresses himself under the electric stimulation: 'There it is. It was like witchcraft. He was in the process of doing this, he snatched something from someone . . . a

stick, or something . . . at the top of the road' (the patient had an epileptic seizure each time he witnessed someone snatching something from someone – under the guise of a childhood memory when he had snatched a stick from a dog). See Delafresneye, *Brain Mechanisms and Consciousness*, 296–97.

15. TN. '*Espions chimiques*'.

16. TN. This is part of a famous remark made by Jean Cocteau: 'The greatest literary masterpiece is no more than an alphabet in disorder'.

17. TN. The Centre National de la Recherche Scientifique (CNRS) is the major state-funded scientific research institute in France, founded in 1939.

18. Anne Anastasi, *Psychological Testing*, 6th ed. (Englewood Cliffs, NJ: Prentice Hall, 1990 [1955]), 169, 255.

19. Cf. Nikolaas Tinbergen, *L'etude de l'instinct*, trans. B. de Zelicourt and F. Bourlieue (Paris: Payot, 1953); and Nikolaas Tinbergen, *Social Behaviour in Animals* (London: Chapman & Hall, 1990).

20. Clifford T. Morgan, *Physiological Psychology* (New York: McGraw-Hill, 1943 [1941]), 144.

21. C. Spearman, *The Nature of Intelligence and the Principles of Cognition* (London: Macmillan, 1923). Specialists in IQ tests today vigorously criticise the '*g* factor' and the notion of general intelligence. But the lack of practical interest in the notion is related precisely to its universality. The *g* or *gamma* factor is present in all living beings and cannot be used to discriminate between them.

22. TN. For Spearman, noegenesis is the inferential capacity that allows for the acquisition of new information through observation and through the combination of what is currently known.

23. Spearman, *Nature of Intelligence*, chap. 4.

24. D. H. Lawrence, *Last Poems*. We borrow both this citation of Lawrence and the next from Leone Vivante's in-depth study in *A Philosophy of Potentiality* (London: Routledge and Kegan Paul, 1955).

Bibliography

Adrian, A. F. *The Physical Background of Perception.* Oxford: Clarendon Press, 1947.

Anastasi, Anne. *Psychological Testing.* Englewood Cliffs, NJ: Prentice Hall, 1990.

Armstrong, Edward A. *La vie amoureuse des oiseaux.* Trans. Jane Fillion. Paris: Albin Michel, 1952.

Ashby, William Ross. *A Design for a Brain.* New York: Wiley & Sons, 1952.

Bachelard, Gaston. *La matérialisme rationnel.* Paris: Presses universitaires de France, 1953.

Benoît, J. 'Etats physiologiques et instinct de reproduction chez des oiseaux'. In *L'instinct dans le comportement des animaux et de l'homme.* Paris: Masson et Cie, 1956, 177–260.

Berkeley, George. *A Treatise Concerning the Principles of Human Knowledge.* Ed. Kenneth Winkler. Indianapolis, IN: Hackett, 1982.

Bernal, J. D. 'The Origin of Life'. *New Biology* 16, no. 12 (1954), 12–18.

Blum, Harold. *Time's Arrow and Evolution.* Princeton, NJ: Princeton University Press, 1951.

Bonner, John Tyler. *Morphogenesis: An Essay on Development.* Princeton, NJ: Princeton University Press, 1952.

———. *Cells and Societies.* Princeton, NJ: Princeton University Press, 1955.

Bourlière, F. *Vie et moeurs des mammifères.* Paris: Payot, 1951.

Bray H. G., and K. White. 'Organisms as Physicochemical Machines'. *New Biology* 16, no. 70 (1954).

Buytendijk, Frederik Jacobus Johannes. 'L'instinct d'alimentation de l'expérience chez crapauds'. *Archives néerlandaises physiologie de l'homme et des animaux* 2 (1918), 217–28.

Dalcq, Albert. *L'oeuf et son dynamism organisateur.* Paris: Albin Michel, 1941.

de Beer, Gavin. 'Embryology and Evolution'. *Journal of Philosophical Studies* 5, no. 19 (1930), 482–84.

———. *Embryos and Ancestors.* Oxford: Clarendon Press, 1940.

Driesch, Hans. *La philosophie de l'organisme.* Trans. M. Kollmann. Paris: Rivière, 1921.

———. 'The Potency of the First Two Cleavage Cells in Echinoderm Development: Experimental Production of Partial and Double Formations'. In *Foundations of Experimental Embryology.* Ed. Benjamin H. Willier and Jane M. Oppenheimer, 38–50. New York: Hafner Press, 1964.

Dupréel, Eugène. *La pragmatologie*. Brussels: Les Editions du Parthenon, 1955.
Eccles, John Carew. *The Neuro-physiological Basis of Man*. Oxford: Oxford University Press, 1953.
Eddington, Arthur Stanley. *The Philosophy of Physical Science*. New York: Macmillan Company, 1939.
———. *Nature of the Physical World*. Cambridge: Cambridge Scholars Press, 2014.
Fauré-Fremiet, Emmanuel. 'Symétrie et polarité chez les ciliés bi- ou multicomposites'. *Biological Bulletin* 79 (1945), 106–50.
Fessard, Alfred. 'Mechanisms of Nervous Integration and Conscious Experience'. In *Brain Mechanisms and Consciousness*. Ed. Jean-François Delafresneye, 200–36. Oxford: Blackwell, 1954.
Field, Mary, J. V. Durden and F. Percy Smith. *See How They Grow: Botany Through the Cinema*. Harmondsworth: Penguin, 1952.
Fontaine, M. 'Analyse expérimentale de l'instinct migrateur des poissons'. In *L'instinct dans le comportement des animaux et de l'homme*. Paris: Masson et Cie, 1956, 151–67.
Frey-Wyssling, Albert. *Submicroscopic Morphology of Protoplasm*. Trans. May Hollander. New York: Elsevier Publishing Company, 1953.
Gamow, George. *The Birth and Death of the Sun*. New York: Viking Press, 1940.
Garstang, Walter. 'The Theory of Recapitulation: A Critical Re-statement of the Biogenetic Law'. *Zoological Journal of the Linnean Society* 32 (1921), 81–101.
Gesell, Arnold. *L'embryologie du comportement*. Paris: Presses universitaires de France, 1953.
Goldschmidt, Richard. *Physiological Genetics*. New York: McGraw Hill, 1938.
Granger, Gilles Gaston. *Méthodologie économique*. Paris: Presses universitaires de France, 1955.
Gudger, Eugene Willis. 'Fishes That Rank Themselves Like Soldiers on Parade'. *Zoologica: Scientific Contributions of the New York Zoological Society* 34 (1949), 99–103.
Guilbaud, Georges Théodoule. *La cybernétique*. Paris: Presses universitaires de France, 1954.
Guiraud, Pierre. *Les charactères statistique du vocabulaire*. Paris: Larousse, 1956.
Haldane, J. B. S. 'Les aspects physico-chimiques des instincts'. In *L'instinct dans le comportement des animaux et de l'homme*. Paris: Masson et Cie, 1956, 454–57.
———. *The Biochemistry of Genetics*. New York: The Macmillan Company, 1954.
———. 'The Origins of Life'. *New Biology* 16 (1954), 12–27.
Hédiger, Heini. *Observations sur la psychologie animale dans les parcs du Congo Belge*. Brussels: Institut des Parcs Nationaux du Congo Belge, 1951.
———. *Les animaux sauvages en captivité: Introduction à la biologie des jardins zoologiques*. Paris: Payot, 1953.
———. 'Instinkt und territorium'. In *L'instinct dans le comportement des animaux et de l'homme*. Paris: Masson et Cie, 1956, 521–41.
Herskovits, Melville J. *Les bases de l'anthropologie culturelle*. Paris: Payot, 1952.
Holtfreter, Johannes. 'A Study of the Mechanism of Gastrulation'. *Journal of Experimental Zoology* 95, no. 2 (1944), 171–212.
Huxley, Julian. *Evolution in Action*. New York: Harper, 1953.

———. 'Evolution as a Process'. In *Evolution as a Process*. Ed. Julian Huxley, A. C. Hardy and E. B. Ford, 1–23. London: George Allen & Unwin, 1954.

———. *Fourmis et termites*. Trans. William Perrenoud. Brussels: Office de publicité, 1955.

Jordan, H. J. 'Indéterminisme vital et le dynamism des structures causales'. *Recherches Philosophique*, vol. 2. Ed. Alexandre Koyré, Henri-Charles Puech and Albert Spaier, 18–47. Paris: Boivin, 1933.

Kuo, Zing-Yang. 'The Genesis of the Cat's Response to the Rat'. *Journal of Comparative Psychology* 11 (1930), 1–35.

Lack, David. *Darwin's Finches*. Cambridge: Cambridge University Press, 1983.

Lashley, Karl. 'The Problem of Serial Order in Behaviour'. In *Cerebral Mechanisms in Behaviour*. New York: Wiley & Sons, 1951, 112–36.

Lehrman, Daniel. 'On the Organisation of Maternal Behaviour'. In *L'instinct dans le comportement des animaux et de l'homme*. Paris: Masson et Cie, 1956, 475–520.

Lichtenberg, Georg Christoph. *Aphorismes*. Trans. Marthe Robert. Paris: Club Français du Livre, 1947.

Lillie, R. S. *General Biology and the Philosophy of Organism*. Chicago: University of Chicago Press, 1945.

Lorenz, Konrad. 'La théorie objectiviste de l'instinct des animaux et de l'homme'. In *L'instinct dans le comportement des animaux et de l'homme*. Paris: Masson et Cie, 1956, 51–76.

———. *King Solomon's Ring*. Trans. Marjorie Kerr Wilson. London: Routledge, 2004. (Orig. pub. 1949.)

Mandelbrot, Benoit. 'Structure formelle des textes et communication'. *Word* 10 (1954), 1–27.

Merleau-Ponty, Maurice. *Adventures of the Dialectic*. Trans. Joseph Bien. Evanston, IL: Northwestern University Press, 1973.

Meyer, François. *Problématique de l'évolution*. Paris: Presses universitaires de France, 1954.

Moles, Abraham. *Théorie de l'information et perception esthétique*. Paris: Flammarion, 1957.

Morand, Pierre. *Aux confins de la vie*. Paris: Masson, 1955.

Morgan, Clifford T. *Physiological Psychology*. New York: McGraw-Hill, 1943.

Morley, D. W. *The Ant World*. London: Penguin, 1953.

Needham, Joseph. *Chemical Embryology*. Cambridge: Cambridge University Press, 1931.

———. *Biochemistry and Morphogenesis*. Cambridge: Cambridge University Press, 1942.

———. *Order and Life*. Cambridge: Cambridge University Press, 2015.

Ombredane, Andre. *L'aphasie et l'elaboration de la pensée explicite*. Paris: Presses universitaires de France, 1951.

Plotinus. *Ennead III*. 1–9. Trans. A. H. Armstrong. Cambridge, MA: Harvard University Press, 1967.

Poirier, René. 'Le problème de l'âme et du corps'. *Société Française de Philosophie* 48, no. 4 (1954), 97–158.

———. 'Henri Poincaré et le problème de la valeur de la science'. *Revue philosophique de la France et de l'etranger* 74, nos. 10–12 (October/December 1954), 485–513.

Portmann, Adolf. *Animal Forms and Patterns: A Study of the Appearance of Animals*. New York: Schocken Books, 1952.

Rostand, Jean. *Les grands courants de la biologie*. Paris: Gallimard, 1951.

Russell, Bertrand. *Introduction to Mathematical Philosophy*. New York: Dover, 1993. (Orig. pub. 1919.)

Russell, Edward Stuart. *Instinctive Behaviour*. London: Edward Arnold, 1934.

———. *The Behaviour of Animals. An Introduction to Its Study*. London: Edward Arnold, 1938.

Ruyer, Raymond. *Éléments de la psycho-biologie*. Paris: Flammarion, 1946.

———. *La cybernétique et l'origine de l'information*. Paris: Flammarion, 1954.

———. 'Les postulats du sélectionnisme'. *Revue Philosophique de la France et de l'etranger* 146 (1956), 318–53.

———. *Neofinalism*. Trans. Alyosha Edlebi. Minneapolis: University of Minnesota Press, 2016.

Schrödinger, Erwin. *What Is Life?* Cambridge: Cambridge University Press, 1948.

Spearman, Charles. *The Nature of Intelligence and the Principles of Cognition*. London: Macmillan, 1923.

Thompson, D'arcy Wentworth. *Growth and Form*. London: Dover, 1942.

Tinbergen, Nickolaas. *L'etude de l'instinct*. Trans. B. de Zelicourt and F. Bourlieue. Paris: Payot, 1953.

———. *Social Behaviour in Animals, with Special Reference to Vertebrates*. London: Chapman & Hall, 1990.

Tomlin, E. W. F. *Living and Knowing*. New York: Faber & Faber, 1955.

Vendryès, Pierre. *Vie et probabilités*. Paris: Albin Michel, 1942.

Vivante, Leone. *A Philosophy of Potentiality*. London: Routledge and Kegan Paul, 1955.

von Frisch, Karl. 'Lernvermögen und Ergebunden Tradition im Leben der Bienen'. In *L'instinct dans le comportement des animaux et de l'homme*. Paris: Masson et Cie, 1956, 345–86.

von Neumann, John. 'The General and Logical Theory of Automata'. In *The World of Mathematics*, vol. 4, 2070–98. London: Dover, 2003.

von Neumann, John, and Oskar Morgenstern. *Theory of Games and Economic Behavior*. Princeton, NJ: Princeton University Press, 2004.

von Uexküll, Jakob. *Animal Forms and Patterns: A Study of Appearances of Animals*. Trans. H. Czech. New York: Schocken Books, 1967.

Waddington, C. H. *Principles of Embryology*. London: George Allen & Unwin, 1956.

Weber, Max. *The Protestant Ethic and the Spirit* of *Capitalism*. Trans. Talcott Parsons. New York: Charles Scribner & Sons, 1958.

Whitehead, Alfred North. *Adventures of Ideas*. New York: Free Press, 1967.

Wiesner, B. P., and N. M. Sheard. *Maternal Behaviour in the Rat*. Oxford: Oliver & Boyd, 1933.

Wilkie, J. S. *The Science of Mind and Brain*. London: Hutchinson's University Library, 1953.
Woodger, Joseph Henry. *Biological Principles*. London: Kegan Paul, 1929.
———. 'The Concept of "Organism" and the Relation Between Embryology and Genetics'. *Quarterly Review of Biology* 6 (1931), 178–207.

Index

active deployment, 9, 41, 93
actualisation, 167–68, 175–76
adaptation, 6, 83, 103, 106–7, 138, 142
adenosine triphosphate (ATP), 34, 43, 46, 182
aggregate, 24, 57, 164
alcohol, 38
allesthetics, 137–38, 140, 142
amboceptor(s), 20, 46, 182
amino acids, 43
Amblystoma, 89, 97
ammonium oleate, 15–16
amoebae, 6, 7, 9, 42, 56, 59, 61–64, 71, 78, 91–92, 143, 145, 160–61, 171–72
amphioxus, 3–4, 10, 78, 88–90, 97, 101, 123, 187n6
anatomy, xiv, xv, 37
Anaxagoras, 154
animism, 30, 38
anti-Darwinism, 123
ants, 66–67, 76, 121, 122
ant-lion, 42
anthropomorphism, 171, 175
anti-finalism, 107
apthous fever, 37
Ariadne's thread, 110
Aristotle, 61
Aron, Raymond, 113
Ashby, W. R., 19, 23, 179
assembly, 3, 12, 19, 123, 145, 192n1; assembly line, 24
assimilation, 51, 161
Atta, 66, 122
atom, 22, 32, 35–36, 38–42, 51, 150, 159–60, 164, 168, 180

auto-conduction, 6–7, 39, 143, 145, 148, 160, 162
automata, 24, 46, 73, 75
automation, viii, 24, 26
Autuori, M., 122
auxin, 81 83, 87, 89, 153

Bacillaria paradoxa, 40
Bachelard, Gaston, 35, 150
Balzac, Honoré de, 118
Baudelaire, Charles, 114, 116
behaviour, 6–9, 26, 38, 43, 67, 73, 75, 88, 113, 115, 121, 123–24, 131, 143, 154, 160, 163–64; formative, 36; free, 4, 78; instinctive, 19, 183; motor, 128; psychological, 19; regulated, 93–112; sexual, 64; structuring, 35, 39; unitary, 160
behaviourism, 96, 156
Bertalanffy, Ludwig von, 46
Benedict, R., 119
benzene, 31, 35, 38–39, 181
Berkeley, George, xiv, 138–39, 149, 162, 191n2
Bergson, Henri, 40, 80, 185n24
Bernal, J. D., 17, 33–34, 36
Bernard, Claude, 148
Binet-Simon test, 172–73
biology, xiv, 14, 29, 36, 62, 64, 121, 124, 159; evolutionary biology, 21; microbiology, 39. *See also* psychobiology
Birch, A., 99
blastula, 4, 6, 15, 130, 164, 178, 193n10

blastopore, 3, 4, 78, 128
Blériot, Louis, 27, 180n43
blood vessels, 90
Blum, Harold F., 181n21
Bolk, Lewis, 24
body, 6, 12, 35, 55, 90, 99, 143, 155, 181, 192; 'projected body', 7
Bohr, Niels, 32, 148
bonds, 15, 29, 30, 33–41, 43, 45–46, 49, 54, 181n15, 184n5, 193n6
Bonner, John Tyler, 18, 40, 59–60, 64, 87, 181n23
Bossuet, Jacques-Bénigne, 131
Boutroux, Emile, 36
brain, xiv–xv, 6–7, 11, 88–89, 103, 106, 143, 146–48, 163, 167, 169, 175
Brillouin, Léon, 46, 50
Brownian motion, 117, 180

Cabanis, Pierre Jean Georges, 154
calcium, 17, 19, 187n5
calculator, 64, 170–71
calliope, 115, 123, 189n8
camouflage, 94
candle, 48
Canguilhem, Georges, 108
cannon, 46
Cannon, B., 148
carbon, 31, 35, 50; carbonic acid, 100; carbonic oxide, 38
catenary, 18, 179n27
causality, 20, 33, 51, 63, 72–73, 120
cells, xiv, 3–4, 9, 14–15, 23, 31–32, 34, 40, 53–62, 66, 68, 74, 78, 89, 90–93, 122, 134, 138, 143–44, 154, 157–59, 164, 171, 177n3
cerebral; activity, 76; analysers, 73; area, 140; centre(s), 109, 142–43, 161; control, 163; consciousness, 164, 174; relays, xiv; zones, 7, 56, 104. See also cortex
Champy, C., 189
chance, 46, 107, 109–10, 113, 117, 122, 173, 175, 180; Markovian chance, 124

chemistry, 30, 31, 36–40, 57, 160, 168; organic, xiii, 35,
child, 20, 81, 101, 145, 172; childhood memory, 194
Chinook (language), 102
chromosomes, 22, 33, 157, 158
Claparède, Édouard, 174
cognac, 38
Cole, K. C., 17
combustion engine, 19, 21
consciousness, 26, 38, 39, 43, 110, 115–16, 120, 147, 149, 150–55, 160–66, 170, 172–73, 175, 193; human, 146; mnemic, 75; perceptual, 139, 144; secondary, 151; 'surveying consciousness', 49
cortex, 7, 46, 102, 103, 142, 161, 193
cosmos, 48
Cournot, Antoine A., 107, 113, 123, 188n25
creation, 1, 12–13, 25–26, 128, 149, 160, 165, 175; artistic, 2; organic, 144–45
Cymbella, 40
crop, 98–99, 109; crop-milk. See prolactine
crystals, 40, 177; crystalline lattices, xviii; liquid crystals, 15–17;
culture, 65, 119, 150; 'biological culture', 65, 173; cell cultures, 163; human culture, 26, 51, 64
Confucius, 119
cybernetics, 18, 96, 171
cytoplasm, 33, 87

D'Annunzio, Gabriele, 101, 187n12
Dalcq, Albert de, 3, 138, 177n1
dam, 46
Darwin, Charles, 21
de Beer, Gavin, 25, 88
de Broglie, Louis, 148
death, 9, 22, 45, 135
Delurance, E. P., 67
Democritus, 170
Descartes, René, 47, 180n8; Cartesian biologists, 30

determination, 10–13, 17, 59, 97, 102, 127, 158–59, 165, 168, 187n6
determinism, 32–33, 72, 159, 165; crypto-determinism, 13; indeterminism, 36
deus ex machina, 161
differentiation, 25, 9–12, 14, 53, 58–61, 74, 88, 90, 128, 159, 163, 164
division, 22, 54, 56–59, 62, 74, 107, 120, 135, 157, 159; cellular, 53, 90; of labour, 64, 67; social, 108
Dodds, E. C., 81
dog, xiii–xv, 98, 135, 157, 172–73, 187n7
domain, xv, 35, 37, 50, 93, 155, 160, 161, 169, 171, 193n6; absolute, 39–41, 138; subdomains, 164; unitary, 55
domanial unity, 45, 49
dominoes, 40, 72–73
Driesch, Hans, 14–16, 55–56, 86, 128
Drosophilia. See fruit fly
dualism, 164
Ducrocq, A., 115, 189n8
ducks, 96–97
Dupréel, Eugè, 64, 159

earthworm, 7, 54, 97
Eccles, J. C., 163
Eddington, Arthur, xiii, 32, 36, 48, 180n8, 182n34
eels, 79, 101
egg, xv, 3–4, 9, 14, 15, 29, 54–56, 63, 68, 71, 72, 78, 95, 99, 103, 106, 108–9, 113, 122, 124, 128–30, 147, 152, 157–59, 167, 169; egg white, 89; egg yolk, 187
Eiffel Tower, 54
electrons, 22, 35, 170, 173, 184n5, 191n4
Éléments de psycho-biologie (Ruyer), 193n11
embryo, 4, 6–12, 15, 21, 36, 50, 54–55, 62, 68, 72, 87, 88, 98, 106, 111, 128, 131, 133, 143, 148, 156, 161, 163, 164, 167, 169, 177n3

embryogenesis, 2, 9, 21, 22, 50, 138, 158, 160, 168, 193n3
embryology, 5, 8, 13, 63, 72, 80, 96, 148, 156; embryology of behaviour, 7
emergence, 1, 26, 44, 150
emotion, 149, 185
encephalitis, 104
energy, 35, 43, 35–51, 101, 164, 170, 182, 184, 192n9, 193n6; free, 128; potential, 12; surface, 17
entelechy, 36, 44, 56–57, 86, 164
entropy, 8, 12, 45–50
enzymes, 23, 31, 153
Epicureans, 140
epidermis, 11, 72, 74, 89, 143; presumptive, 10
epigenesis, 57, 101, 129, 158, 174
equilibrium, 4, 8, 11, 18–19, 23, 47–48, 112, 118, 121, 170; disequilibrium, 79
equipotentiality, 55–57, 160, 174
evolution, 21–23, 25–26, 30, 33–36, 44–47, 81–82, 96, 121, 125, 131, 135, 137, 146, 158, 173; linguistic, 123; social, 120
existentialism, 131
external circuit, 19, 42, 144, 151, 155
eye, 10, 17, 18, 20, 30, 31, 54, 88, 90, 123, 127, 130, 137, 139, 140, 142, 143–48, 152, 179n31, 184n3

faradic current, 7, 167
Fauré-Fremiet, Emmanuel, 16
feedback, 19, 20, 21, 27, 45–46, 48, 98, 100, 109, 112–13, 138, 163
fibrous structure, 128
fighting fish, 97
film, xiv, 117, 162. *See also* microfilm
finalist activity, 171
finality, 47, 171, 172
finches, 151, 192n5
fingernails, 10
fishermen, 71, 81
flies, 74
flowers, 54, 124, 137–38, 159
Foraminifera, 17

force, 17, 18, 55, 118, 166, 176, 193n6;
  of gravity, 73, 183, 183n14
Fourier, Joseph, 66–67, 183n14
fox, 71
freedom, 13, 32–38, 42, 54, 131
Freudian, 104
Frey-Wyssling, Albert, 30–31, 36, 41, 50
fruit fly, 38, 124

Gamow, George, 38, 48, 181n20, 182n34
Garstang, Walter, 25
gastrula, 3–4, 6, 11, 15, 85, 147, 179n18
gastrulation, 3–4, 10, 68, 78, 90, 131, 158–59, 164, 193n3
genes, xiii, 22–24, 33, 38, 44, 53, 108, 124, 130, 157–58, 167
genetics, 25, 36, 124, 158, 160
Gesell, Arnold, 7, 62, 128
Gestalt theory 14, 18, 96, 101, 159
Gewächs, 115
Gide, André, 20
Giersberg, M. F., 88
giraffe, 121
glycogen, 20
gradients, 18, 60, 79, 129
grafts, 5, 10–13, 88–90, 97, 104, 123, 134, 158–59
Grassé, Pierre-Paul, 108
Gray, J., 101
great Argus pheasants, 142, 144
god, 148, 176
Goldschmidt, Richard, 124, 131
gravity, 18, 73, 78, 183
Guilbaud, G., 114, 120
Guyénot, E., 123

habit, 56, 103, 111, 192n5
Haldane, J. B. S., 27, 33–34, 36, 76, 100
hands, xv, 6, 23, 33, 48–49, 54–55, 143–44, 161, 175, 177
harmony, 21, 31, 63
Hartman, Carl, 7
Harvey, E. N., 17, 148
Hebb, Donald, 96, 106–7, 110

Hédiger, Heini, 108, 154–56
Hegel, G. W. F., 119, 120, 123
Heinroth, Oskar, 96
Heitler, Walter, 39
heredity, 7, 37, 101
Hiroshima, 22–23
history, 34, 42, 44, 107, 113, 120–25, 127, 130–31
Holtfreter, J., 78, 87, 91–92, 134
Holst. *See* von Holst
homeostat, 19, 23, 46, 99, 100, 109, 112, 179
homeostasis, 26, 27, 47
homo oeconomicus, 121
Homer, 119
homunculus, 6–7, 33, 38
hormones, 24, 71, 81, 83, 87–88, 90, 99, 100, 107, 134, 152–54
Hull, Clark, 96
humanity, 119, 149
Husserl, Edmund, xiv, 161–62
Huxley, Julian, 26, 82, 138
hydra, 16, 59, 101, 122
hydrogen, 31, 37, 40, 48

ideal, 65, 110, 131, 161; cause, 180; form, 164; sense, 169; theme, 167; vague ideal, 109
iodine, 89
imitation, 51, 64, 71
in-itself, xiii
inductor, 13, 24, 72, 74–75, 81, 87–91, 123, 128, 132–35, 153–56
information, 2, 12, 13, 50–51, 73–76, 80, 116, 153–54, 165, 184n9; machines, xiv
intelligence, 40, 172–75
invention, 25, 67, 111, 129, 162, 171, 174–75; technical, 21
iodine, 89
isomorphism, xiv, xv, 54

jackdaw, 76, 185
Jackson, Hughlings, 102, 155, 163, 192n3

Jordan, H. J., 14–15, 32, 36
Joyce, James, 117–18

Kant, Immanuel, xiv
Kardiner, Abram, 119
keratin, 33
'key atoms', 33, 180n2
kidney, 16, 18, 54
knowledge, xiii–xv, 1, 40, 64, 73–74, 86, 148, 150, 155–56, 160, 174
Köhler, Wilhelm, 46, 48, 110, 130, 184n13
Kuo, Z. Y., 107, 188n24

La Fontaine, Jean de, 112
Lack, David, 151, 192n5
lake, 47–48
Lao Tzu, 95
Lashley, Karl Spencer, 102–3
Lawrence, D. H., 176
learning, 56, 96, 101, 106–7
Lehrman, Daniel, 96, 99, 103, 106–10
Leibniz, Gottfried Wilhelm, 120, 131, 162, 190
Lichtenberg, Georg, 44, 48
life, 8, 23, 32–34, 36, 39–45, 68, 69, 74, 112–13, 119, 128, 154, 168–69, 174; cultural, 51; economic, 121; family, 64; cycle, 61; organic, 19, 79
Lillie, R. S., 165
limbs, xiv, 7, 16
limb buds, 11, 89–90, 97, 123, 161
Lissmann, H. W., 101
lithium, 108
Loeb, L., 134
Logos, 120, 176
London, Fritz, 39
Lorenz, Edward, 64–65, 82, 96–97, 100–106, 109
lungs, 9
Lüscher, Max, 71

machine gun, 19
magnesium, 108
Mallarmé, Stéphane, 114

Mandelbrot, Benoît, 76, 107, 116–18
marble, 12–13, 172–73
marble block, 23, 26, 68
Markov chains, 113–18, 120, 122, 124, 127, 130, 132–35
Markov jargon, 113–34
mathematics, xiv
Maxwell, James, 116
meaning, 75, 94, 105–8, 127
meccano, 29, 34, 37, 40
mechanics, 32, 35, 127
melody, 93, 95, 101, 105, 112, 115, 161; melody-duration, 39; mnemic melody, 109, 160; releaser melody, 183; 'vertical' melody, 77
membrane, 8, 37, 45–46, 48, 150
memory, 21, 25, 39, 56, 79, 91, 166, 168–69, 193–94n14; psychological, 167; reproductive, 105; transspatial, 131
mendelian, 96
Merleau-Ponty, Maurice, 120, 189n14
metaphysics, 174
Meyer, François, 48, 120
microfilm, 8, 9
microorganisms, 32, 46, 49–50
microphysics, 32, 49
Microstomum linearum, 122
mnemic potential, 74
mnemic themes, 79, 115, 131, 168–69
molecules, 10, 15, 17–18, 31, 33–34, 38–43, 46, 51, 57, 117, 150, 160, 164, 168, 180
Moles, A., 80
mollusc, 93, 155
Montaigne, Michel de, 173
Morgan, C. T., 173
morphology, xiii–xiv, 27, 31–37, 44, 109, 122, 133, 155
mousetrap, 19
multiplicity, 35, 63, 150
muscles, xiv, 7, 9, 20, 62, 78, 101–3
mutation, 21, 22–26, 82, 123–24, 158, 171

Mystery of Picasso, The (film), 161–62
mythology, 2, 29, 149

narcolepsy, 104
natural selection, 26, 82, 108, 138
nature, xv, 26, 30, 107, 142, 170, 177
Needham, John, 11, 47, 86–87, 134
neogenesis, 174–75, 194n22
nerve cells, 9, 78
nervous system, xiv–xv, 4, 6, 7, 9–10, 12, 17, 19–20, 26, 39, 75, 82, 90, 92, 97–98, 100–101, 105–6, 133, 144, 148, 160, 163–64, 166–67, 170, 184
neurologists, 102–4, 151
nucleus, 22, 82, 157, 158
nitrogen, 42
non-being, 176

occipital zone, 6, 145
ontogeny, 25
opossum, 7
organic forms, 30–34, 36, 40, 53, 57, 151, 169–70
organisation, 3, 15, 23, 26, 29, 41, 47, 48, 50, 66, 102, 103, 107, 123, 134, 138, 156, 158, 164, 167; psycho-organic, 33–34; social, 120
organisin, 71, 184n3
organogenesis, 175
organs, 6, 9, 14, 15, 21, 42, 46, 47, 53, 56, 57, 73, 74, 95, 122, 133, 137, 138, 142–45, 149, 151, 155, 170, 182; decorative, 123; genital, 99; proto-organs, 17; sensory, 78, 150
osmosis, 46

painting, 144, 159, 161–62
pancreas, 20, 46
panpsychism, 36
pantograph, 14, 15, 19, 73, 184n6
Papilio centralis, 124
Papilio sphyrus, 124
parallelism, 7

Parmenidean, 176
Parsons, E. C., 119
partridges, 71
patterns, 30, 94, 124, 138
Pelobates fuscus, 89
peptides, 43
perception, xiv, 40, 56, 67, 73, 75, 86, 111, 137–45, 148–54, 160, 185, 192
Peters, H. M., 156
philosophy, 61, 108, 145; German romantic, 151; mechanist, 41; philosophy of, 122, 131
phonograph, 106, 166
photosynthesis, 43, 170
photon, 140, 184; photonic radiation, 138
phylogeny, 25
physics, 34, 39, 40–42, 48, 157, 164, 184n5; classical (mechanistic), 14, 32, 33, 45, 72, 170; indeterminist, 36; quantum, 181n20
physiogenesis, 35
physiology, xiv–xv, 35, 37, 45, 97, 109; psycho-, 128
Piaget, Jean, 111
Planck, Max, 41
Plato, xiv
Platonism, 96, 123, 164, 174
Plessner, Helmuth, 115
Plotinus, xv, 177n4
Polistes, 67
Portmann, A., 94, 138
preadaptation,
preformationism, 29, 87, 106, 109, 129, 131, 157
Prigione, Ilya, 51
primordia, 3, 4, 7–8, 10–11, 13–16, 20, 49–50, 54, 63, 90, 127–33, 139, 161
prolactine, 81, 98
proteins, 33, 37, 40, 42, 43, 45–46, 86, 134–35, 145, 164, 167–68, 174, 185
protozoa, 6, 39, 42, 54, 56, 59, 143–44, 170

providentialism, 67, 131
pseudo-formations, 1–3
pseudopods, 6, 7, 143
psychology, 19, 61–62, 72, 75, 86, 151, 163, 171, 183n14, 187n10

quantum chemistry, 35–36
Queneau, Raymond, 102

Rabelais, François, 66, 183
Racine, Jean, 119
radiation, 22–23, 48, 108, 138
Radiozoa, 17
Rana arvalis, 89
Rana esculenta, 97
realisation, 170, 175
reason, 170–73
recollection, 105, 110
red-breasted robin, 71, 80
reflex, 8, 83, 101–2, 111, 156
regeneration, 8, 56, 59,61, 78, 123–24, 184n3, 186n9
Reiss, David, 99
reproduction, 22, 25, 37–38, 44, 122, 174; internal, 59–63; self-, 24, 33, 53–57
resemblance, 68
resonance, 35, 37, 120, 132
retina, 140, 143, 145, 153
Richter, Curt, 96, 99
Rignano, Eugenio, 128
Ringer's solution, 97, 187n4
Romantics, the, 36
Rostand, Jean, 29
Russell, Bertrand, xiii
Russell, E. S., 71, 82, 86, 98, 110

Sagittaria, 14
salt, 57, 99–100, 109
Scarron, Paul, xiii, 177n3
Schopenhauer, Arthur, 192n9
Schrödinger, E., 25
Schwind, Joseph, 90
seagull, 64
sea urchin, 14, 54

sex, 24, 34, 64, 68, 82, 87, 94, 100, 108–9, 149, 151–54, 191
skeleton, 17, 35, 41
sketch. *See* primordium
Shannon, Claude, 116–17
snails, 20, 179n31
soap bubble, 15, 17–19
social institution, 65
sociology, 19, 57, 64, 66
space-time, 154–55, 164, 168
sparrowhawk, 115
spearman, 130, 174
species, 5, 13, 21, 23, 24–25, 29, 33, 65, 75, 79, 82, 88, 89, 91, 104, 107–10, 113, 120, 122–27, 131–32, 138, 143, 149, 151, 152, 155, 158; 'species prejudice'; 173
Spemann, 87–88, 133, 156
spicules, 17, 179n24
spiders, 42, 74–75, 82, 86, 93, 155–56
spiritual, 164
sponges, 17, 58
starlings, 97–98, 152
Staudinger, Hermann, 36
steam engine, 19
stickleback, 64, 71, 75, 78, 80, 183n1
sugar, 19–20, 46
super constellation, 27
survey, 19, 41, 45, 46, 49, 115

tadpole, 90, 97
technology, 171, 174
telephone, 12, 74, 76, 176
television, 105, 140, 174, 191
termites, 71, 108
Tetramorium atratulum, 122
thematism, 14–15, 18, 25, 42, 63, 115, 151, 163; non-spatial (trans-spatial); thematism, 109, 111, 164, 174
thermodynamics, 49
thermostat, 19
Thimann, K. V., 153
thyroxin, 89–90
Tinbergen, Nikolaas, 64, 66, 71, 76, 82, 96, 100–106, 110, 183n1

Titus Livius, 114, 120, 189n4
tobacco mosaic virus, 37
Trembley, Abraham, 59
tubes, 45, 191n4, 192n1
Twitty, Victor, 89, 91–92
tympan, 11, 72, 74, 81

Umanski, K. B., 91–92
Umgebung, 154
Umwelt, 113, 150, 154–55
unconscious, 151, 185n18
universe, 105 147, 150, 170
Urodela, 88
Urostyla, 16
utilitarian, 65, 94, 144, 149
utopia, 66–67, 183n9

Valéry, Paul, 119
Vendryès, Joseph, 81, 102, 117
Venus, 1
Venus flytrap, 42, 73–74
vertebrae, 138
Viaud, Julien, 77
viruses 22, 25, 29, 32–33, 37–40, 42–44, 49, 51, 53, 134, 143, 160, 174, 181n18
vitalism, 30, 36, 43; neo-vitalism, 55
von Holst, Erich, 97, 101, 103
von Neumann, John, 24, 25, 53, 121, 190n18

von Uexküll, Jakob J., 94, 154
volvox, 58–62

Waddington, C. H., 11, 15, 85–86
wasps, 67
water, 17, 19, 38–39, 42–43, 46–49, 57, 96, 100, 103; 'hot water' and 'cold water', 80; water-mill, 46
Watson, D. L., 128
Watt, H. J., 86, 186n2
Watt, James, 19
wave mechanics, 38
wear, 8, 19, 43, 47–48
Weber, Max, 120
webs, 42, 74–75, 82, 86, 93, 120, 155
Weiss, Pierre, 97, 187n4
Whitehead, Alfred North, 42, 65, 165
Whitman, Charles, 86
wings, 20–21, 97, 123–24, 142, 172, 175
Witt, P. N., 156
Wolff, Étienne, 191n7
Woodger, Joseph Henry, 57–58, 159
Würzburg school, 132, 151, 163

Yahweh, 30

Zilla x notate, 156
Zipf, George, 116–19, 189n9
zone of 'turning', 31

About the Author and Translators

Raymond Ruyer (1902–1987) was an influential philosopher of science and Professor of Philosophy at the Universite de Nancy.

Jon Roffe is Lecturer of Philosophy at Deakin University and the Melbourne School of Continental Philosophy, and a founding editor of *Parrhesia: A Journal of Critical Philosophy*. He is the co-editor or author of a number of volumes on twentieth-century French philosophy, including *Badiou's Deleuze* (2012).

Nicholas B. de Weydenthal is a graduate from the School of Historical and Philosophical Studies at the University of Melbourne. He wrote a doctoral dissertation on risk, environmental management, organizational theory and philosophy.

www.ingramcontent.com/pod-product-compliance
Lightning Source LLC
Chambersburg PA
CBHW020124240426
43673CB00038B/581